THE FLYING SAUCERS & THE US AIR FORCE
THE OFFICIAL AIR FORCE STORY

BY
LAWRENCE J. TACKER

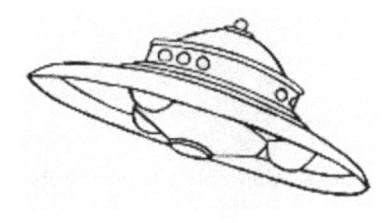

SAUCERIAN PUBLISHER

ISBN: **978-1-955087-24-7**

2022, Saucerian Publisher

Al rights reserved. No part of this publication maybe reproduced, translate, store in a retrieval system, or transmitted in any form or by any means, electronic, mechanical, photocopying, recording or otherwise, without prior written permision from the publisher.

PROLOGUE

It is generally a good idea to return to the classics in any genre. This also goes for UFO literature. Rereading a book, or reviewing old documents after ten or twenty years is a rewarding experience. You will discover new data and ideas you didn't notice before. The reason, of course, is that you are, in many ways, not the same person reading the book the second or third time. Hopefully you have advanced in knowledge, experience, intellectual and spiritual discernment. A good starting point is to reread the contactee classics material of the 1960s, in order to understand the deeper mystery involved in what happened during that era.

The 1960 book entitled: *Flying Saucers And The U S Air Force*, by Lt. Colonel Lawrence J. Tacker was "anti-UFO," but it is essential because it contains many items of interest, including a collection of letters from civilians and a reproduction of Air Force UFO regulations and press releases. The author was the official Air Force spokesman on UFOs in 1960 at the Pentagon4. This book was published at the same time as Donald Keyhoe's *Flying saucers: Top secret* (1960) and Keyhoe is mentioned several times.

This book is a facsimile reproduction of the original printed text in shades of gray. Because this material is culturally important, we have made it available as part of our commitment to protect, preserve and promote knowledge in the world. This book has been formatted from their original version for publication. **IMPORTANT, although we have attempted to maintain the integrity of the issues accurately, the present reproduction could have missing and blurred pages poor pictures due to the age of the original scanned copy.** Because this material is culturally important, we have made it available as part of our commitment to protect, preserve and promote knowledge in the world.

Editor
Saucerian Publisher

FLYING SAUCERS
and the U. S. Air Force

Veteran of 15 combat missions and two battle campaigns as an 8th Air Force Navigator in the European Theater of Operations, Lieutenant Colonel Lawrence J. Tacker, 43 years old, of Falls Church, Virginia, is-now assigned as-Chief ©f-the Magazine-arid Book Branch, Office of-lafarmotion-, Qfficeof the Secretary of the Air Force, Washington 25, D. C.

Colonel Lawrence Tacker with approximately 4,000 flying hours is rated as a Master Navigator with the United States Air Force and is presently fully qualified in the latest Air Force multi-jet engine aircraft. He holds the Air Medal, the Purple Heart, the European Theater of Operations Medal with two Battle Stars, the Army and the Air Force Commendation Ribbons and has 19 years service with the USAF.

He, his wife, Dorothy and their four children: Francine Mary, age 14; Thomas Wood, 12; James Lawrence, 10; and Timothy Francis, 7 reside in an old colonial house on two acres of ground in historical and beautiful northern Virginia. By necessity, as well as inclination, Colonel Tacker's hobby interests are remodelling and gardening.

High speed jet aircraft in a fog-shrouded sky sweeps low over the heads of spectators at an air demonstration and gives the startling illusion of being a flying saucer.

FLYING SAUCERS
and the U. S. Air Force

by

Lt. Colonel LAWRENCE J. TACKER

FOREWORD

By an Act of Congress the United States Air Force is charged with the Air Defense of the United States. Rapid identification of anything that flies is an important part of air defense. Thus the Air Force initiated and continues the unidentified, flying object program. Under this program all unidentified flying object sightings are investigated in meticulous detail by Air Force personnel and qualified scientific consultants. So far, not a single bit of material evidence of the existence of spaceships has been found.

THOMAS D. WHITE
Chief of Staff
United States Air Force

CONTENTS

	PAGE
Foreword	v
By General Thomas D. White, Chief of Staff, USAF	
Introduction	vii

CHAPTER 1
    THE 1959 PACIFIC SIGHTINGS   3

CHAPTER 2
    THE HISTORY OF THE "SAUCERS"   12

CHAPTER 3
    SAUCERS IN THE NEWS   19

CHAPTER 4
    THE PSYCHOLOGY OF SAUCERS   30

CHAPTER 5
    IT'S EASY TO BE FOOLED   51

CHAPTER 6
    LISTEN TO THE EXPERTS   75

CHAPTER 7
    THE OFFICIAL AIR FORCE POSITION   82

APPENDICES
    1. AIR FORCE REGULATION 200-2   91
    2. USAF TECHNICAL INFORMATION SHEET   99

3. JANAP 146—COMMUNICATION INSTRUCTIONS FOR
   REPORTED SIGHTINGS   112
4. PRESS RELEASE—October 25, 1955   136
5. PRESS RELEASE—November 5, 1957   138
6. PRESS RELEASE—November 15, 1957   144
7. PRESS RELEASE—October 6, 1958   146
8. PRESS RELEASE—January 22, 1959   149
9. PRESS RELEASE—July 15, 1959   151
10. PRESS RELEASE—January 29, 1960   153
11. PRESS RELEASE—July 21, 1960   157

INDEX   163

List of Illustrations

	PAGE
A Jet Aircraft appears to be a Saucer	*Frontispiece*
1. Chart of Sightings in July 1959 Near Hawaii	9
2. A Hoax	48
3. A Lenticular Cloud	52
4. Illusion Produced by Lighting Conditions	52
5. High Altitude Balloon in a Strong Wind	54
6. An Illusion	55
7. A Radar Equipped Lockheed Super Constellation	56
8. A Mirage	58
9. A Fireball	62
10. Radar Phenomena Due to Abnormal Conditions	66

FLYING SAUCERS
and the U. S. Air Force

1
THE 1959 PACIFIC SIGHTINGS

It was a beautiful Pacific night. Clear skies prevailed overhead and the stars looked as though you could reach up and touch them. Fleecy white clouds visible below were scattered so that occasionally the sheen from the broad Pacific could be seen by the pilot of the DC-7 Pan American airliner flying at 20,000 feet.

Captain C. A. Wilson carefully checked his watch. 1300 Zebra he thought. That's 3:00 A.M. Honolulu time. About four more hours and Pan American Airways Flight #947 out of San Francisco would be safely on the ground at Honolulu International Airport. He turned slightly in his seat so that he could see his co-pilot, Richard Lorenzen, and his flight engineer, Robert Scott. The co-pilot had just adjusted the aileron trim and was studying his instrument panel. The flight engineer had just completed checking his instruments and was leaning forward in his seat looking toward Captain Wilson. Wilson checked his position report again: 28 degrees, 25 minutes North; 144 degrees, 30 minutes West, all instruments in the green; just another routine run for Pan Am Flight #947. In the rear most of the passengers dozed; a few read and the two flight stewardesses were ready to attend to any of their needs.

And suddenly it happened!

Wilson couldn't believe his eyes. Lorenzen and Scott stared open mouthed at the spectacle they saw. It seemed to be about 1,000 feet above the airliner and slightly to the left—a mysterious cluster of bright white lights moving constantly at high speed raced across the heavens from the southwest toward the south, their path, 180 degrees to the flight path of the DC-7. For at least ten seconds they maintained their course and then suddenly seemed to make a sharp right turn at a speed inconceivable for any aircraft known and then just as quickly as they came the lights suddenly disappeared.

The veteran pilot automatically checked his watch, 1302Z (Greenwich Time). His hands automatically went to the radio panel overhead and he tuned in Honolulu Radio Control Center.

On the ground at Honolulu the Air Traffic Controller heard Pan Am

Flight #947 calling in. He responded quickly and received Captain Wilson's reporting position: 28 degrees, 25 minutes North, 144 degrees, 30 minutes West, altitude 20,000 feet; "have just sighted bright white lights moving at high speed in a generally eastern direction, they appeared to be one bright center light with four smaller lights on the left side. The object moved approximately 180 degrees to the aircraft's flight path and made a 90 degree turn. The lights seemed to be slightly higher in altitude than the aircraft. Objects or objects observed and verified by two other crew members. End of message."

At the Flight Control Center all bedlam seemed to break loose. Immediately following Captain Wilson's report, Pan American Airways Flight #942, pilot Captain E. G. Mathwig called in as Captain Wilson finished with his position report for 1302Z.

"Honolulu Control—Honolulu Control. This is Pan Am Flight #942, Position at 1302Z; 26 degrees North, 146 degrees, 58 minutes West, 19,000 feet; observed light in clear sky; sky bright with moonlight; lights were twice the brightness of the planet Venus; object looked like an extremely bright star; appeared in the West and disappeared in the West above the aircraft; possibly a shooting star; no estimate possible of specific distance or altitude; duration of sighting 10 to 15 seconds."

"Honolulu Control—This is Slick Airways, Flight #719, Captain Zedwick, position report 1302Z; 13,000 feet; 26 degrees, 5 minutes North, 143 degrees, 30 minutes West; have just observed bright lights; clear night; no moon; lights appeared as bright as automobile headlights one mile away; color of object pure white and it changed brightness; object appeared to be one large light with four smaller lights in trail; object appeared to come straight at our aircraft and looked like a large trailer or Very Pistol Flare; it appeared in the Southwest and disappeared in the Southwest; flight path seemed level with our aircraft and speed very high, estimate over 1,000 knots; duration of sighting 3 to 5 seconds; it is possible it could have been a meteor."

And again—"Honolulu Control, Honolulu Control. This is Empress Flight 323, Captain L. C. Moffatt, 1302Z position report; 29 degrees, 40 minutes North, 150 degrees, 40 minutes West, altitude 11,000 feet; have just observed strange object; it appeared to be one large light surrounded by a cluster of 6 or 7 smaller lights; largest light-size of a dime; lights were bright as automobile headlights one block away and the color was orange-yellow. The lights moved from the Southwest to the Southeast, moving faster than any known aircraft; duration of sighting 5 seconds; clear skies with trace of daylight; definitely not a meteor or shooting star, no trail visible."

And still another. "Honolulu Control—Honolulu Control. This is United Flight #21, Position 1302Z; 920 Nautical miles East of Honolulu, 22 degrees, 30 minutes North, 142 degrees, 30 minutes West; altitude 12,000 feet; have just sighted white light dead ahead and above aircraft; descended toward aircraft and below; then banked to left. As the object moved away there were four white lights in a rectangle with a large bright light in the center. End of report."

Tense excitement took over. Telephones jangled and duty officers at Headquarters, Commander-in-Chief, Pacific; Headquarters, Pacific Air Forces and Western Sea Frontier were notified. The Air Traffic Controllers called back each flight individually and verified the information they had received from each aircraft and asked aircraft captains to report to Flight Operations immediately upon arrival at Honolulu International Airport.

Headquarters, Pacific Air Forces immediately dispatched intelligence officers to interrogate each of the crews as they arrived to try and determine just what it was they had seen.

Alert aircraft were not scrambled because of the distances involved and the reported direction of the bright object seen by the commercial airliners. "Scrambled" is the flyers' term for a quick takeoff to identify an unknown object or track. Approximately five hours later, between 0715 and 0900 local time, intelligence officers interrogated all five of the crews who had reported seeing the bright object. In addition, four other crews, Canadian Pacific Flight 323, Pan American Flight #945, Slick Airways Flight #601 and Pan American Airways Flight #752 all reported that they had seen a bright object in the Southwest which was evidently the same object reported by the original five crews.

As soon as the crews' interrogation had been completed and their statements taken, the information was sent immediately to the Aerospace Technical Intelligence Center at Wright-Patterson Air Force Base, Ohio for analysis and evaluation.

Newspaper reporters, of course, had their day and talked to each and every airline crew member arriving in Honolulu that morning. The wire services carried the story quickly to all parts of the world and the Honolulu Star Bulletin, the Los Angeles Times, the Milwaukee Journal, the Baltimore Sun and many others published headline stories on these impressive sightings. "Pilot Reports Unidentified Flying Object," said one newspaper;

"High Fast Objects Sighted Over the Pacific by Airline Pilots" headlined another. And so it went, in the Sunday newspapers of July 12, 1959. Under each and every headline, the stories stressed the fact that these were veteran airline pilots who had seen mysterious unidentified flying objects, flying at speeds "faster than anything I've ever seen." A few of the pilots said the objects seemed to bear down on their aircraft but turned away abruptly at the last minute and then the lights disappeared. According to one newspaper account, Captain Wilson made the following statement: "We were cruising at 20,000 feet, with low clouds decked below us when the object first appeared about 1,000 feet above us and to our left," he said. "My co-pilot and flight engineer both saw the bright light as it came toward us at an extremely high rate of speed. For at least ten seconds it maintained its course which was on an opposite heading to us. Had it been another aircraft it would have passed well to our left.

"Suddenly the object made a sharp right turn at a speed inconceivable for any vehicle we know of, and the light suddenly disappeared. The smaller lights were evenly spaced and were either part of the mysterious object or this was an example of darn good formation flying." Captain Wilson appeared visibly shaken said the newspaper account. He had never seen anything like this in his 19 years of flying. He added that he had never believed such foreign objects existed. "I'm a believer now," he said.

Captain Noble Sprunger, another Pan American Airways pilot, who waited until landing to make his report said the object looked something like a falling star or meteor. "The only thing startling about it was that it appeared to be coming right on track toward us," Sprunger commented.

At Headquarters, United States Air Force in Washington, D. C. the Air Force Press Desk duty officers were busy throughout the week end answering inquiries on these simultaneous sightings. The initial statement given by the Air Force stated that a tentative analysis indicated the object was probably a meteor and that when final evaluation of the sighting had been accomplished by the Aerospace Technical Intelligence Center, the results would be released to the press immediately.

Meanwhile, Aerospace Technical Intelligence Center personnel were busy. Urgent messages from Headquarters Pacific Air Forces had been flashed to the Aerospace Technical Intelligence Center at Wright-Patterson Air Force Base, Ohio and to Headquarters United States Air Force.

Aerospace Technical Intelligence Center personnel checked immediately with the Commander, Western Sea Frontier, Headquarters Air Research and

The 1959 Pacific Sightings

Development Command, Headquarters 28th Air Division at Hamilton Air Force Base, California, with Navy Liaison offices and the Missile Division of the Intelligence Center to make sure that no missiles or satellites known to the United States could have been the cause of these sightings. Air Weather Service was contacted and the 1110th Balloon Activities Group at Lowry Air Force Base, Colorado.

At this point a possible lead was found. There had been a balloon launched from the Vernalis, California Launching Site in the area at 1300Z on 11 July 1959, position approximately 35° north and between 150° and 158° west. These balloons carried bright red lights which are visible for approximately five miles. However, due to this extreme northerly position, the intense brightness of the sighted object, the color difference and the great speed of the sighted object, the possibility of it being a balloon was ruled out.

Captain Mathwig, Pan American Flight #942, had reported bright moonlight. However, a check revealed the moon had set at 2238 local time, or 0838Z, four hours and 24 minutes prior to the sighting. It is possible that the object shed so much light that the Captain was of the impression that the moon was still up. Captain Mathwig also reported that the object was approximately twice as bright as Venus, which at its brightest has a stellar magnitude of −4.4, which is bright enough to cause shadows at night. The object, if approximately twice as bright, would have a stellar magnitude of approximately −5.

Captain Moffatt of Empress Flight 323 reported a trace of daylight. A check revealed that morning daylight began at 0331 local, or 1331Z, 29 minutes after the sighting. Here again it is believed that the extreme brightness of the object was responsible for the misimpression that the light was from morning twilight or daybreak. Captain Moffatt also stated that the object was definitely not a meteor or shooting star because there was no trail visible.

Any meteor which is brighter than −3 magnitude is defined as a bolide or fireball. The color range varies, from a bluish white through white to yellow-green and reddish. Usually when fireballs are seen there is a brilliant flash of light which lasts as long as the meteor is visible but which may fluctuate in brightness. Sometimes these meteors break into several parts. Some meteors leave trails and some do not.

From a report entitled "Long Enduring Meteor Trains and Fireball Orbits" by Charles P. Olivier, Professor Emeritus of Astronomy, University of Pennsylvania, it was determined that of 33,000 meteors observed by Professor Guno Hoffmeister of Sonneburg Observatory, Germany, only 42, or one of every 786, left trails which persisted ten seconds or more. The American Meteor Society indicates a ratio of one in 750; also, of 102 fireballs listed in the report, only 28 had trails which were significant enough to record. According to Dr. Olivier's report, fireballs occur approximately one in every 258 meteors. Fireballs are sporadic and not associated with any particular shower.

The average speed of the 102 fireballs from Dr. Olivier's report is 28 miles per second, approximately 100,800 miles per hour, or 87,480 knots. Curved or irregular paths may be due to irregular shapes. Fireballs sometimes explode and on occasion they may be heard.

After checking with many Armed Forces radar stations and civilian observatories, the facts surrounding these particular sightings were presented by the analyst at the Aerospace Technical Intelligence Center to qualified scientific personnel and to civilian consultants for their opinion regarding the final evaluation for these sightings.

The final conclusion of the Aerospace Technical Intelligence Center stated that in the opinion of the Center the object responsible for the sightings on July 11, 1959 between California and Hawaii was that type of exceptionally bright and large meteor classified as a fireball. In addition, the Center stated it believed that all the reports on 11 July 1959 referred to the same meteor. The time, description and general direction of the object from all witnesses tended to substantiate this opinion.

The Aerospace Technical Intelligence Center evaluation for these sightings was released to the press on 14 July 1959 and confirmed the tentative analysis which had been made on 12 July 1959, one day after the sighting. However, even with this quick result, it did not stop the letters and inquiries from the general public, among them several Congressional inquiries which were sent to the Department of Defense from Congressional committees or individual members of Congress asking for an immediate explanation of these sightings. The Congressional inquiries were, of course, engendered by constituents who asked their duly elected representatives to obtain an explanation. Many of the letters were sober evaluations offering suggestions and expressing concern over what this bright object viewed by "qualified observers" could be. Many of the letters were contemptuous and abusive, hurling allegations and charges against the Air Force and the Department of Defense for withholding vital information from the general

The 1959 Pacific Sightings

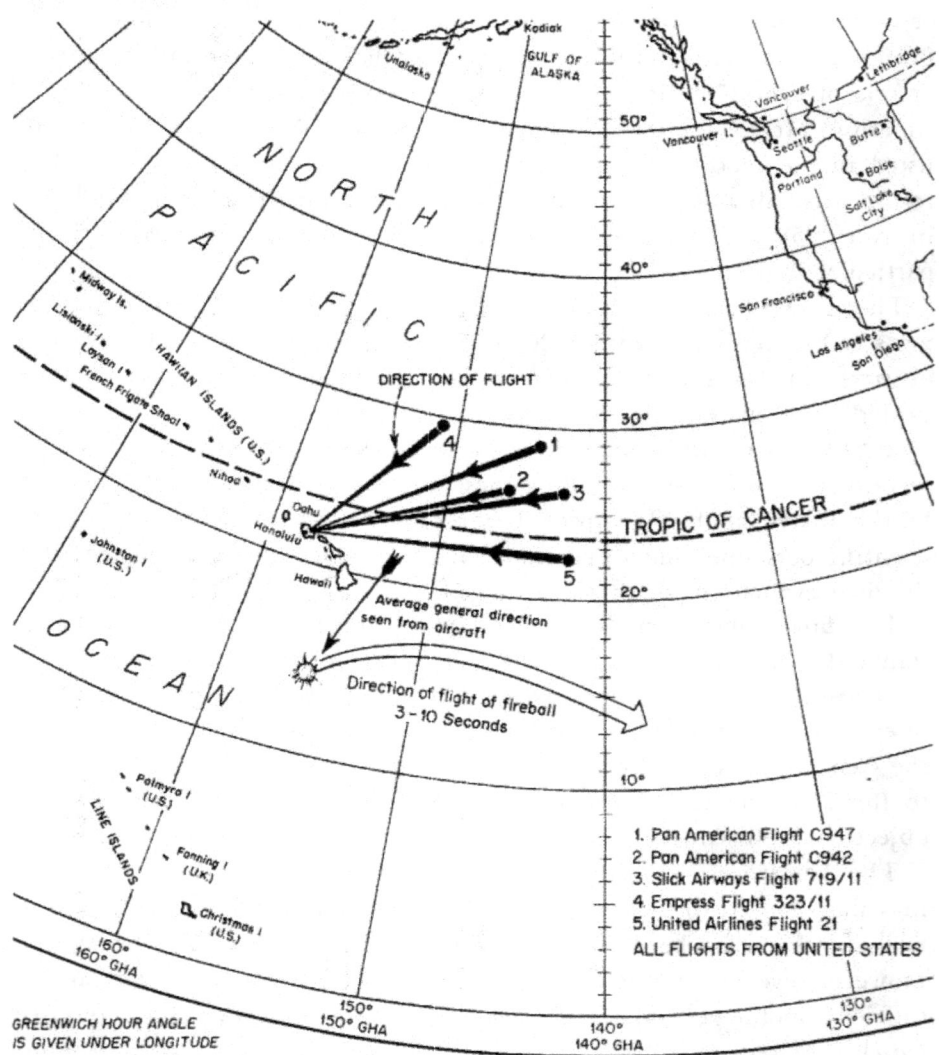

Chart of sightings in July 1959 near Hawaii. 1. Pan Am Flight C947; 2. Pan Am Flight C942; 3. Slick Airways Flight 719/11; 4. Empress Flight 323/11; 5. United Airlines Flight 21.

public which proved that unidentified flying objects or popularly termed flying saucers were spaceships and were seen almost daily in our atmosphere. Most of these letters again stressed the fact that pilots had viewed these latest sightings and referred to previous sightings where members of the Ground Observers Corps, Radar Operators, Navigators and responsible citizens had seen objects which were referred to in the press as possible spaceships.*

There were two recent reported sightings in the Nome-Unalakleet, Alaska area which were very similar to the 1959 Pacific sightings. The first was on February 14, 1960 and the second was on March 6, 1960. An investigation concerning the February 14 sighting developed the following information.

An object described as blunt and rounded, bright in appearance, emitting sparks intermittently, and accompanied by a blue flame was seen from Nome and Unalakleet at approximately four o'clock p.m. The Aerospace Technical Intelligence Center (ATIC), Wright-Patterson Air Force Base, Ohio, concluded that this object was probably a very bright meteor.

Of all the information provided to ATIC concerning a UFO sighting, the most reliable is usually the general description (not to include size), direction from the observer, approximate angle of elevation, duration of the sighting (for very short periods), and the time of the sighting. Using only these factors, it was possible, by trigonometric methods, to determine that the object was approximately 100 miles high. The description of this object, the duration of the sighting, and calculation of the 100 mile altitude led to the conclusion that this object was probably a meteor.

Subsequent to the ATIC conclusion on this case, Dr. Christian Elvey, Director of Geophysics Institute, Alaska University, concluded that the object in question was a very bright meteor. Other possibilities, such as missiles, satellites, et cetera, were considered but ruled out due to inconsistencies. All of the times reported were consistent to within five minutes, an error in time that can normally be expected. The following information was developed as a result of the investigation concerning the March 6 sighting.

A UFO was reported on March 6 in the same general area between Nome and Unalakleet, Alaska. This report originated with an airline pilot who was flying at about 4,500 foot altitude, 86 miles south of Nome. The object was described as being a bright flash seeming to result from an explosion behind the aircraft. The pilot and co-pilot reported that the flash was so

---

*The Pacific sightings of 11 July 1959 were purposely selected for the opening chapter because of the highly trained personnel who reported these sightings to the United States Air Force. Some poetic license was taken in attempting to dramatize this incident. Essentially though, the facts are as reported. This well known case proves the fact that anyone can see an unidentified flying object under extenuating circumstances and can be justly mystified for a period of time at least, by what they have seen.

The 1959 Pacific Sightings

bright that they were both blinded for approximately two seconds. However, they heard no sound. The only passenger aboard the aircraft was asleep, and neither heard nor saw anything. Two individuals on the ground, a 14 year old girl in Nome, and the town marshal of Nome, reported seeing a flash.

There is a difference of opinion among witnesses as to whether an explosion or loud noise accompanied the UFO. ATIC has concluded that this object was a very bright fireball. The only difference is that a loud explosion or report is associated with the bolide type meteor.

And so it goes from day to day. Reports from all over the world are referred to the United States Air Force for evaluation and accompanying the reports are the ever-present charges that the Air Force knows that space travel in reverse exists, that flying saucers from other planets or galaxies are visiting our atmosphere and that the Air Force refuses to divulge this information to the general public because they would panic and chaos would result. As we will see these charges are ridiculous.

2

THE HISTORY OF THE "SAUCERS"

Even the ancient Romans saw flying saucers but referred to them as flying shields. In the writings of Julius Obsequiens, a 4th Century Latin Poet, he referred to a phenomena he observed near Rome in the year 100 B.C., described as follows: "At sunset a circular object like a shield was seen to sweep across the sky from West to East."

Among other sightings, in the year 1554 a mysterious object flashed across the sky near St. Chamis, France and, in the year 1709 many strange phenomena invaded the skies of Europe. On November 23, 1896 at Oakland, California a strange cigar shaped object appeared in the sky and was seen by thousands of people as it sailed eastward, moving slowly and creating newspaper headlines from California to Chicago where it disappeared and was never heard from again. Contemporary astronomers identified the object as Alpha Orionis but public opinion was that the object was an airship.

During the 1920s and 1930s, there were sporadic reports of strange objects or phenomena seen in the heavens. But no one referred to them as spaceships or suggested that they came from another planet or star.

World War II and the Korean War gave us a few sightings of strange phenomena or unidentified objects in the sky from pilots flying at higher altitudes and speeds than man had attained up to that time. Our flying personnel jokingly referred to these strange objects or sightings as "foo fighters" or "gremlins." The St. Elmo's fire phenomena, or static electricity, was blamed at the time for most of these sightings.

For all practical purposes, and for the purpose of this book, Air Force history relating to unidentified flying objects or flying saucers began on Tuesday, June 24, 1947. On that day a Boise, Idaho businessman named Kenneth Arnold was flying his private plane in the vicinity of Washington's Mount Rainier. Arnold, a veteran pilot who had logged thousands of hours in the Pacific Northwest, was assisting in a search mission for a C-46 Marine transport plane which had been reported missing. While flying near the jagged peaks of Mount Rainier, Arnold spotted what looked like a chain of nine saucer-like objects playing tag among the peaks. Upon landing

The History of the "Saucers"

Arnold immediately reported his sighting and this report set off a chain reaction which has not stopped even to this day.

Within a few days after his report, the flying saucer had spun into the national spotlight. Observers from all over the country reported seeing flying saucers, flying tear-drops, flying lights, and flying pie plates. Reports poured in to the United States Air Force through its many air bases and were referred to the United States Air Force's Materiel Command at Wright-Patterson Air Force Base, Ohio. Exhaustive investigations for each reported sighting were launched and the Air Force's interest, or program, was born or, one might say, it evolved from necessity.

This project was to have a number of names as the Air Force progressed in its investigation of sightings of unidentified flying objects up to the present date. In the beginning, reported sightings of so-called flying saucers or flying disks were sent to the Air Force's Air Materiel Command for analysis and evaluation. Letters, telegrams, and telephone calls were initially exchanged among Headquarters Air Materiel Command, Headquarters USAF, Washington, D. C. and the other major commands, such as the Air Training Command, the Air Defense Command, the Tactical Air Command and the Strategic Air Command.

This was a new and perplexing problem for the Air Force, and it presented many difficulties in setting up an investigative organization and a standard operating procedure to cope with the many reports that were coming in.

In September, 1947, the Air Materiel Command stated that "it is possible within the present day United States knowledge, provided extensive detailed development is undertaken, to construct a piloted aircraft which has the general description of the unidentified flying objects being reported." The Air Materiel Command went on to say that any developments in this country along the lines indicated would be extremely expensive, time-consuming, and would be undertaken at considerable expense to present day projects. Then, based upon its limited experience in this field, the Command recommended that the United States Air Force issue a directive assigning a priority, security classification, and code name for a detailed study of this matter, to include the preparation of complete sets of all available and pertinent data which would then be made available to other Interested Governmental agencies.

In December, 1947, Headquarters United States Air Force stated that it would be Air Force policy *not* to ignore reports of unidentified flying object sightings in the atmosphere and that the Air Force recognizes that part of its mission is to collect, correlate, evaluate, and act on information of such a nature.

In implementing this policy, the Air Force directed that the activities of the project should include proper reporting procedures and the project was assigned the code name of "Sign" with a security classification of restricted for military use. The official birthday for Project Sign was January 22, 1948. Subsequently the project name changed to "Grudge" and then finally became "Project Bluebook," under the auspices of the Aerospace Technical Intelligence Center at Wright-Patterson Air Force Base (ATIC), a name it has retained up until the present time. Their statistical records are available at all times to accredited newspaper reporters and other media representatives.

Although occasional sightings of strange aerial objects were reported prior to Mr. Arnold's sighting of June, 1947, it was this sighting that touched off the flying saucer era we know today. Kenneth Arnold, private pilot and representative of a Fire Control Equipment firm in Boise, Idaho, was en route from Chehalis, Washington to Yakima, Washington on an errand of mercy in a privately-owned plane when he first saw the saucer-like craft approaching Mount Rainier.

"I could see their outline quite plainly against the snow as they approached the Mountain," he reported. "They flew very close to the mountain tops, directly south to southwest, down the hog-back of the range, flying like geese in a diagonally chain-like line as if they were linked together."

Arnold's observations included the fact that the objects seemed much smaller than a DC-4 but he judged their span to be as wide as the outboard engines on either side of a DC-4's fuselage. "They were approximately 20-25 miles away and I couldn't see a tail on them," he declared. "I watched for about three minutes, a chain of saucer-like things, at least five miles long, swerving in and out of the high mountain peaks. They were flat like a pie pan and so shiney they reflected the sun like a mirror. I never saw anything so fast," he said.

Arnold said he estimated the saucers' speed at about 1,200 miles an hour.

Arnold's story when it broke in the newspapers was treated mainly with amusement and disbelief. Resentful with what he termed press ridicule, Arnold's answer to the newspaper stories was "they can call me Einstein, Flash Gordon, or just a screwball, but I'm absolutely certain of what I saw." He added that if he ever again saw a phenomenon in the sky, even if it were a ten story building flying through the air, he would not say a word about it.

Today no one knows just what Arnold did see on Mount Rainier. The objects were judged to be of non-astronomical origin according to a report submitted under the original saucer project. The non-astronomical origin was determined by Professor Joseph A. Hynek, prominent Astrophysicist and then head of the Ohio State University Observatory. Dr. Hynek worked under contract with the Air Force's Air Materiel Command on an intelligence investigation of flying saucer sightings to determine what per cent might definitely be attributed to astronomical phenomena. In his review of the Arnold incident, Dr. Hynek touched upon certain inconsistencies in the Arnold estimates of size, speed, and performance of the saucers. It appears probable, Hynek reported, "that whatever objects were observed were travelling at subsonic speeds and may, therefore, have been some sort of known aircraft."

In the days that followed Arnold's observation, the number of disk or saucer sightings reported to the Air Force began to snowball. At Muroc, California a group of Air Force officers reported spotting spherical objects of a disk-like shape whirling through the skies at a speed in excess of 300 miles per hour. In Portland, Oregon several businessmen told investigators they saw a group of disks that wiggled, disappeared and reappeared several times. Only a few days after Arnold's sighting a saucer was reported seen over his hometown of Boise. "A half circle in shape, clinging to a cloud and just as bright and silvery looking as a mirror caught in the rays of sun."

The sightings continued to come in and fantastic details were associated with the phenomena reported.

In January, 1948, the Air Force established its initial standard operating procedures for reporting, investigation, analysis and evaluation of reported sightings.

As with all new projects early methods of analysis and evaluation were not complete due to the unknown nature of the reported phenomena. One cannot say that errors were made but rather that through inexperience omissions were probably made in not covering every possibility. Reporting agencies were impressed with the necessity for getting all the factual evidence on sightings, such as photographs, physical evidence, radar sightings, and data on size and shape. Persons sighting such objects were encouraged to engage the assistance of others when possible to get more definite data. For example, military pilots were told to notify neighboring air bases by radio of the presence and direction of flight of an unidentified flying object so that other observers in flight or on the ground could assist in its identification. Many sightings by qualified and reliable witnesses had been reported. However, each incident seemed to have unsatisfactory facts associated with it, such as shortness of time under observation, inaccurate estimates of distance from the observer, the vagueness of descriptions, or lack of photographs, inconsistency between individual observers and a serious lack of descriptive data that prevented definite conclusions being drawn. Explanation of some of the incidents revealed the existence of simple and easily understandable causes so that there appeared initially the possibility that enough incidents could be solved to eliminate or greatly reduce the aura of mystery already associated with the sightings.

While the Air Force worked toward developing proper and adequate reporting, investigation, analysis and evaluation techniques and procedures, they were charged by some individuals and groups of gross negligence in investigating reported sightings of unidentified flying objects and with actually having evidence of the fact that flying saucers were actually spaceships from other planets or stars.

The Air Force grappled with the many problems involved. Detailed check lists compiled by technical personnel indicating basic elements of information necessary for analysis of each individual sightings were prepared and distributed to appropriate Government agencies. Graphic methods were applied so as to present basic data in such a form that overall facts would be made apparent. Prepared graphical data included maps and charts, locations of air bases, airlanes, locations of radio beacon stations, weather stations, projected celestial phenomena charts and flight paths of migratory birds. The psychological aspects of these sightings were taken into consideration. Members of the Scientific Advisory Board to the Chief of Staff, United States Air Force, provided consultant services on this subject. The Weather Bureau, the Federal Bureau of Investigation, and the Rand Corporation, a civilian research organization under contract to the Air Force, were all engaged in the initial effort to provide a standardized approach to this problem.

Consideration was given to the possibility that the unidentified objects represented scientific developments beyond the level of knowledge attained in this country. However, since at that time the U.S. was probably the most advanced of the industrial nations on the earth, and our interest in scientific developments throughout the world was very active, it would have been necessary for any other country to continue research and development work in extreme secrecy for any such project to have reached such an advanced stage of development without a hint of its existence becoming known.

Another possibility considered was the fact that these aerial objects might possibly be visitors from another planet or solar system. Little is known of

The History of the "Saucers"

the probability of life on other planets so there was no basis on which to judge the possibility that a civilization far in advance of ours existed outside the earth. Various people have suggested that an advanced race may have been visiting earth from Mars or Venus at intervals. The reports of objects in the sky have been handed down through many generations. If there were a race of such knowledge and power, they would surely have established some form of direct contact before now.

The Air Force took this problem seriously and put a lot of effort into developing adequate and proper reporting, investigating, analysis, and evaluation procedures.

Early press releases dealt with specific cases and did not present an overall Air Force position on the subject of unidentified flying object sightings. These early press releases, or notices for the press, issued during the period 1948 through 1954 were not kept for the record and so are not available for this book.

On July 29, 1952 the Air Force issued a press release and held a press conference in the Pentagon, Washington, D. C. on this subject. On January 17, 1953 a panel of scientific consultants to the United States Air Force concluded that the evidence presented on unidentified flying objects showed no indication that these phenomena constitute a direct physical threat to the national security. The panel recommended that the Air Force take immediate steps to strip the unidentified flying objects of the special status and the aura of mystery they had unfortunately acquired.

On October 25, 1955 the Air Force released its study on unidentified flying objects called Project Bluebook, and then Secretary of the Air Force Donald A. Quarles stated "on the basis of this study we believe that no object such as popularly described as flying saucers, has overflown the United States." (See Appendix Four.) On November 5, 1957 in response to inquiries as to results of previous investigations into unidentified flying object reports, the Air Force said that after ten years of investigation and evaluation of UFOs, no evidence had been discovered to confirm the existence of so-called flying saucers. (See Appendix Five) On October 6, 1958 the United States Air Force announced that nothing had ever been found to substantiate any claims that flying saucers were interplanetary spaceships. (See Appendix Seven) On January 22, 1959 the United States Air Force announced that the Air Force UFO study showed the unexplained sightings had decreased to less than 1% of the total reported. (See Appendix Eight) On February 17, 1959 an unidentified flying object policy meeting was held by the Air Force in the Pentagon and recommendations were made to stress accurate and timely press releases for the general public on the subject of UFOs. On July 15, 1959, the United States Air Force reported that UFO sightings for the first six months of 1959 had shown a 50% decrease and on January 22, 1960 the Air Force announced that only 326 sightings had been reported during the entire year of 1959. (See Appendices Nine and Ten.) On July 21, 1960 the Air Force reaffirmed its position concerning unidentified flying objects. (See Appendix E.)

At the present time the selected qualified scientists, engineers and other technical personnel involved in the Aerospace Technical Intelligence Center's analysis of unidentified flying object sightings are bringing Project Bluebook up-to-date to keep the American public informed on this subject. This revised report will probably be published sometime during early 1961. And still the sightings continue to come in, some easily explained, others containing insufficient data to allow for a valid conclusion and a very few with what seems like sufficient data to allow a valid conclusion remain unexplained.

3
Saucers in the News

In addition to the Pacific sightings of July, 1959 mentioned in Chapter 1, there have been over the past 13 years a number of sightings which have caught the imagination of the general public and because some of these sightings remain unexplained or the conclusions of the Aerospace Technical Intelligence Center are unacceptable to avid saucer fans, they are still a topic of conversation whenever and wherever saucer discussions develop.

Some of these sightings could be referred to as the "UFO Classics." A few of these should be discussed for the record so that a proper mood can be established for the next chapter dealing with the psychology of the saucer sightings and the fact that even when objective valid conclusions are offered by the Air Force they are rejected by the confirmed saucer believer.

Tragedy struck during a highly publicized saucer sighting in the year 1948. On January 7th of that year, an unidentified object that looked like an ice cream cone topped with red was sighted near Godman Air Force Base, Ft. Knox, Kentucky by both military and civilian observers. The Godman tower requested a flight of National Guard F-51 Mustang aircraft flying in the vicinity to investigate the phenomenon. Three of the planes closed in on the object and reported it looked to be metallic and was of tremendous size. One pilot, described it as round like a tear drop and with an appearance of being fluid.

The flight leader, Captain Thomas F. Mantell, contacted the Godman Tower with an initial report that the object seemed to be travelling at half of his speed at 12 o'clock high.

"I'm closing in now to take a good look," he radioed. "It is directly ahead of me and still moving about one-half my speed. The thing looks metallic and it is very large. It is going up now and forward as fast as I am—that is 360 miles per hour," Captain Mantell reported from his aircraft. "I'm going up to 20,000 feet and if I am no closer I'll abandon the chase." That was the last radio contact made by Mantell with the Godman Air Force Base tower.

Five minutes after Mantell disappeared from his formation, the two remaining planes returned to Godman Air Force Base. A short time later one aircraft resumed the search, covering territory 100 miles to the South and as high as 33,000 feet but found nothing.

Subsequent investigation by the Accident and Investigation Board and the medical examiner revealed that Captain Mantell had lost consciousness between 20,000 and 30,000 feet from lack of oxygen and had died of suffocation before his plane crashed.

The mysterious object which the flyer chased to his death was first incorrectly identified as the Planet Venus. However, later investigation revealed that in January, 1948 at the time of the Mantell crash, large skyhook balloons were being launched from Clinton County Air Force Base in Southern Ohio. Weather maps also indicated that prevailing winds would have carried the balloons westward as they climbed and the jet stream would have taken the balloons to a position south or southwest of Godman Air Force Base.

The final conclusion of the Aerospace Technical Intelligence Center was that Captain Mantell had chased such a balloon on that fatal day. Strong wind currents holding the balloon in an almost horizontal position could have given the fluid tear drop shape as described by the pilots. This was probably the case. In any event, Mantell did die either chasing a skyhook balloon or another possibility, as suggested by Dr. Donald H. Menzel of the Harvard Observatory, Captain Mantell may have mistaken some form of aerial phenomena such as a mock sun for a flying saucer, flown too high without oxygen and crashed.

Perhaps the most fantastic saucer sighting in the Aerospace Technical Intelligence Center's record was the widely publicized spaceship which two Eastern Airlines pilots, Captain C. S. Chiles and 1st officer John B. Whitted reported encountering in the sky near Montgomery, Alabama on July 24, 1948.

In the sky near Montgomery (on the same day) presumably the same object was seen by ground observers at Robins Air Force Base, Marietta, Georgia about an hour earlier. All reports agree that it was travelling in a southerly direction trailing varied colored flames and that it behaved like a normal aircraft insofar as its disappearance from the line of sight was concerned.

The Eastern Airlines pilots, Captain C. S. Chiles and 1st Officer John B. Whitted, described the phenomenon as a wingless aircraft, 100 feet long, cigar-shaped and about twice the diameter of a B-29 Superfortress with no protruding surfaces. "We saw it at the same time and asked each other 'what in the world is this?'" Chiles told investigators. "Whatever it was, it flashed down toward us and we veered to the left. It veered to its left and passed us about 700 feet to our right and a little above us. Then, as if the pilot had seen us and wanted to avoid us, it pulled up with a tremendous burst of flame from the rear and zoomed into the clouds, its prop wash or jet wash rocking our DC-3 aircraft."

The flame-shooting mystery craft as described by the Eastern Airlines pilots had no fins, but appeared to have a snout similar to a radar pole in front, and gave the impression of a cabin with windows above.

Captain Chiles declared the cabin "appeared like a pilot compartment, except brighter." He said the illumination inside the body itself approximated the brilliance of a magnesium flare.

"We saw no occupants," he told investigators. "From the side of the craft, came an intense, fairly dark blue glow that ran the entire length of the fuselage, like a blue flourescent factory light. The exhaust was a red-orange flame, with a lighter color predominant around the outer edges."

The pilots said the flame extended 30 to 50 feet behind the object and became deeper in intensity as the craft pulled up into a cloud. Its speed was estimated to be about $1/3$ faster than common jet aircraft.

In their investigation of this prominent sighting, the Aerospace Technical Intelligence Center personnel screened 225 civilian and military flights scheduled and found that the only other aircraft in the vicinity of Montgomery, Alabama at the time was an Air Force C-47. Application of scientific analysis to the incident indicated that a fuselage of the dimensions reported by Chiles and Whitted could support a load comparable to the weight of an aircraft of this size at flying speeds in the subsonic range.

This object is still carried in Air Force files as unexplained.

About noon on Sunday, July 20, 1952, the United States Air Force Press Desk in the Pentagon received the following report from the Senior Controller on duty at the Air Route Traffic Control Center of the then Civil Aeronautics Authority, National Airport, Washington, D. C.

At 11:40 PM, EDT, on July 19, 1952, Air Route Traffic Control radar operators picked up from seven to ten unidentified images on their radar scopes at the National Airport. The unidentified images appeared to be in the vicinity of Andrews Air Force Base, Maryland and seemed to be travelling at approximately 100 to 130 miles per hour. The Air Route Traffic Control Center advised Andrews Air Force Base and the Military Flight

Service Center at Middletown, Pennsylvania, and a remote radar center some 200 miles from Andrews Air Force Base in Maryland.

According to the controller and later confirmed by Air Force sources, Andrews Air Force Base radar operators were unable to pick up these images on their radar scopes.

At approximately 3:15 AM, EDT, 20 July 1952, a pilot of a Capital Airlines flight, out-bound from National Airport, reported sighting several lights between Washington, D. C. and Martinsburg, West Virginia. They were described as moving rapidly up and down and horizontally, as well as hovering in one position. Shortly after that the pilot of a Capital-National Airlines flight reported that a light had followed him from Herndon, Virginia to within four miles of touchdown at Washington National Airport. This information was relayed to the proper Air Force agencies, including the Air Force Intelligence Section in the Pentagon and the Aerospace Technical Intelligence Center at Wright-Patterson Air Force Base, Dayton, Ohio.

The solution for these sightings was relatively simple and was quickly found.

The radar and visual sightings were due to a temperature inversion at the time. This is an abnormal condition wherein a layer of warm air overlays a cooler air mass and a duct is formed through which radar pulses travel and reflect ground targets from great distances. Radar pulses normally travel in a straight line and therefore are limited in range for picking up surface targets due to both the earth's curvature as well as signal strength, and that ducting causes the signal to follow the earth's curvature, therefore allowing for returns from a surface target at greater than normal distances. The temperature inversion also explains the visual sightings because with a layer of warm air over cool air the path of light rays is lengthened to parallel the earth's surface at greater distances and this condition in many instances may cause a visual mirage.

In another case, in the Spring of 1957, a commercial airliner flying between New York and San Juan, Puerto Rico, at 19,000 feet, sighted a large bright object that appeared to be coming directly toward the airliner. It was described by the pilot as a magnesium-flash with a pale green tint. The pilot swerved his aircraft violently to avoid collision with the strange aerial object.

The Aerospace Technical Intelligence Center checked all military flights in the Atlantic for that date and determined that there were no unusual

Saucers in the News

planes or missile activities for that day. In addition, they found that five of the commercial aircraft flights in approximately the same area reported similar sightings. Position plots on the map showed that all planes had sighted the same object. Three of the aircraft reported the object seemed to split apart in the air. The description, of course, coincides with the known features of a bolide or fireball. As a final check all data for this sighting was submitted to the Smithsonian Astrophysical Observatory and they confirmed and concurred with the Aerospace Technical Intelligence Center's conclusions.

There's a marked similarity between this case in 1957 and the airlines sighting of July 1959 in the Pacific, discussed in Chapter 1.

On 24 February 1959 at 8:45 PM, the pilot of an American Airlines Flight in the vicinity of Bradford, Pennsylvania enroute from Newark Air Field, New Jersey to Detroit, Michigan, sighted a mysterious lighted object and called the attention of his passengers to this strange sight. This sighting was also seen by a United Airlines pilot who reported seeing the strange lights 50 miles East of Youngstown, Ohio at 8:45 PM. Ground observers also reported the strange lights over Akron, Ohio at 9:15 PM. The pilot of the American Airlines Flight made the following statement:

"It was approximately 2045. I noticed these three lights off my left wing in the vicinity of Bradford, Pennsylvania. I was flying at 8,500 feet on top of broken clouds. Visibility was unlimited with no upper clouds observed. It was extremely difficult to ascertain the distance of the lights. The color of the lights were from a yellow to a light orange. The intensity of the lights also changed from dim to a bright brilliance. Sometimes the interval of the three lights were identical to the Belt in the Constellation Orion. Occasionally the rear lights lagged somewhat behind. Also they changed altitudes. During the 40 minutes of observation, the three lights occasionally came forward from a 9 o'clock position. Also occasionally the light extinguished completely alternating from one to another, sometimes the whole three were extinguished and during this whole operation, as I mentioned before, the lights changed in intensity. This action was not only seen by myself but four crew members and passengers on board and also by two other airplanes in the area.

"The only possible explanation other than flying saucers could be a jet tanker refueling operation. Never having witnessed a refueling operation at night, I am not aware of the lighting of the jet tanker.

"My air speed during this complete flight was 250 knots indicated. I also do not know the air speed of tankers during operation if this could be so. I contacted Air Traffic Control to find out if they had any airplanes on a clearance and no three airplanes were given.

"In summary, it was difficult for me to believe they were jets because of low speed and configuration. If they weren't jets I still don't know anymore than I did before even though I watched them for 40 minutes before. Due to the dark and strong lights I was not able to ascertain any size or shape. The altitude of the objects was 30 degrees above my horizon. Distance away is unknown."

The geographical area concerned was bordered on the North along the New York-Pennsylvania border by the route of American Airlines Flight #139, which had departed Newark, New Jersey at 7:10 PM and was scheduled to arrive at Detroit, Michigan at 10:52 PM. On the South the geographical area concerned was bordered by the Pittsburgh, Pennsylvania—Akron, Ohio locale, overflown by the United Airlines flight reporting this incident. The American Airlines pilot, Captain Killian said "sometimes the interval of the three lights were identical to the Belt in the Constellation Orion." This was initially mentioned as a possible solution by the United States Air Force, with a qualifying statement that the report as submitted had not yet been analyzed and that the findings of the Aerospace Technical Intelligence Center would be based upon a complete analysis and evaluation of the reported sighting.

The pilot's written statement suggested the possibility that he had witnessed a night aerial refueling operation. This statement was furnished to the Air Force by Mr. J. A. Maxwell, Manager of Operations, American Air Lines, Detroit, Michigan and was taken from the pilot's flight report turned in to the airline operations office immediately after the flight.

This sighting turned out to be B-47 type aircraft accomplishing night refueling from a KC-97 tanker aircraft. The pilot's report confirmed this, and Air Force records indicate that three B-47 aircraft were in the geographical area mentioned on night refueling operations. Air Force KC-97 tanker aircraft have several groups of lights which at a distance would appear to be one or more lights. The time duration of the refueling operation varies and can last well over an hour, depending on the type of operation.

The KC-97 tanker refueling a B-47 aircraft normally flies at an altitude of approximately 17,000 feet, at approximately 230 knots true air speed. This accounted for the lights seen by Captain Killian being approximately 30 degrees above his aircraft and remaining in view for approximately 40

Saucers in the News

minutes. These facts also coincide with his report of low speed and general configuration of the object or objects.

In addition, since the tanker was making a ground speed of approximately 210 knots (230 knots true air speed, with a 20 knot headwind) and the United Airlines pilot first reported seeing the lights at 8:45 PM, 50 miles East of Youngstown, Ohio heading toward Akron, Ohio, a distance of 120 miles, this also accounts for the tanker aircraft lights being sighted over Akron at 9:15 by ground observers. The final proof was supplied by the 772nd Aircraft Control and Warning Squadron at Claysburg, Pennsylvania which confirmed the fact that three B-47 type aircraft were conducting night refueling operations in the area.

Here are a few recent less well-known sightings with Aerospace Technical Intelligence Center's findings.

At approximately 3:00 AM, September 21, 1958, a woman in a small northern Ohio town was awakened by a very bright light which illuminated her bedroom. Looking out of a bedroom window which faced her front yard, she saw an object which was flat and circular with a dome-shaped top. The witness reported the object as six to eight feet from the ground, moving in a northerly direction parallel with the length of the house, descending with a floating oscillating motion. The object was reported as approximately ten feet from the window when first sighted. The witness insisted that the top of the object was clear to her in every detail, that the color reminded her of dull aluminum, and the dimensions of the object were estimated to be a 20-foot diameter and a six foot thickness. The item was reported to have made several turns around the yard and then rose instantly out of sight. Duration of the sighting was estimated as five minutes.

The Air Force investigation revealed that a railroad track ran near the home of the witness. Contact with a railroad official revealed that a train passed the home of the witness at approximately the hour and date of the sighting. This train has an oscillating headlight and this light could have been seen by the witness. Contact was also made with the local Coast Guard Station and the Officer-in-Charge reported that he was using his spotlight on the Coast Guard Cutter to attract the attention of another ship, and that his light was directed toward shore in the general direction of the witness' home.

Time and date of this incident was coincident with those reported by the witness. The weather at the time of the sighting, as determined from the Coast Guard log, was intermittent mist and rain with haze and smoke. The Air Force conclusion is that a combination of lights coupled with the prevailing weather was responsible for the illusion experienced by this witness.

At 4:25 AM on the morning of September 29, 1958, an object generally described as round, half again as large as the moon, colored a glowing orange, with sparks shooting to the rear, and making a humming sound as of rushing wind was sighted by a number of witnesses between Washington, D. C. and Pittsburgh, Pennsylvania. All witnesses reported the object at approximately the same time. No more than a minute's difference was reflected in any of the reports. The reports also had the same general description and the same direction of motion. In the Washington area, several witnesses reported that the object, after flying overhead, landed and a brilliant white light could be seen on the ground. Investigation by the Air Force determined that a light on a dairy farmer's barn, which was being used again after a long period of inoperation, was responsible for the illusion that the object had landed. As to the object itself; its description, speed, and other general characteristics point to its being a fireball. In passing overhead its course took it over the dairy farm and the horizon. The bright light on the barn created the illusion that it had stopped and landed. The Air Force conclusion in this case is that the object responsible for this sighting was a fireball.

Two civilians from a mid-Eastern state were driving near a dam on the evening of October 26, 1958. When they rounded a curve approximately two to three hundred yards from a bridge they saw what appeared to be a large, flat, egg-shaped object hovering approximately 100 to 150 feet above the bridge superstructure. They slowed their car and when they got to within 75 to 80 feet of the bridge their engine quit and the lights of the car went out. The driver stopped the car. Attempts were made to re-start the car and when this was unsuccessful, they became frightened and abandoned the automobile. Putting the car between themselves and the object, the witnesses watched for approximately 35 to 45 seconds. The object then seemed to flash a brilliant white light, and both men felt heat on their faces. This was followed by a loud noise, and the object began rising vertically. While

Saucers in the News

rising the object became very bright, and its shape could not be determined. It disappeared in five to ten seconds. After the object disappeared the car was started. The witnesses turned and drove to the nearest telephone where they contacted the local police. Two patrolmen were sent to investigate the call, and the men told them of their experience. The witnesses then noticed a burning sensation on their faces, and became concerned about possible radiation burns. Both men went to a city hospital for an examination, and both were advised by the doctor that they were not burned and had no reason for concern. An extensive investigation was made of this incident. However, no valid conclusion could be found from available information as to the possible cause of the sighting, and it remains unexplained.

On March 13, 1959, at 6:29 PM, an unidentified flying object was sighted in a northern mid-western state. This sighting was reported both visually and on radar by military witnesses, airborne and on the ground. The object as seen visually was described as round, and ranging in color from red to green. Only one object was reported, it having a high speed, straight flight path, and it disappeared by fading from sight. All witnesses agreed that the object had a magnetic bearing of approximately 300 degrees. When this object was sighted interceptor aircraft of the Air Defense Command were scrambled, and headed toward the object at top speed without any noticeable rate of closure. Unsuccessful at their attempts to intercept, the fighter aircraft abandoned the mission.

This case was thoroughly investigated by the Air Force. Every witness was interviewed, the possibility of air traffic in the area was checked, as well as weather and any other unusual occurrence which could account for the sighting. Radarscope film taken during the hour of the sighting was obtained and analyzed. The analysis of the radar film revealed that the radar readings were due to interference and not the existence of a real target. As to the visual sighting, there was every indication that this object was the planet Venus. One witness reported that he felt the object was a star or a planet. Venus at the time of the sighting was just on the horizon, and was visible only due to refraction. The apparent color change and motion was due to alternate layers of air with different temperatures. The fact that the reported object maintained its same relative position, with the same azimuth and elevation as Venus at that time and date, that high speed interceptors were unable to effect closure, and its disappearance by fading from sight substantiates the conclusion that the object was Venus.

At approximately 1:30 AM on the morning of March 22, 1959, a young married couple motoring in a northwestern state sighted an unidentified flying object. They described the object as dome shaped, 20 to 30 feet in diameter, intensely lighted, silverish, with a flashing red light, two pale windows, and converging shafts of light shining from ports on the underside of the object. When first sighted, the object was to the southwest at what was estimated at one and one half miles at approximately 200 feet altitude. It was reported as hovering. However, as the auto carrying the witnesses approached, it appeared to parallel their path 50 to 75 feet south of the road on which they were travelling in an easterly direction. After approximately eight to ten minutes the object rose rapidly and disappeared. To add to the eeriness of this sighting, there was no sound associated with the object at any time. The couple who reported this were determined to be of above average intelligence and of excellent character. The witnesses were disturbed over publicity which was being given to their report. Investigation of this case ruled out most of the common causes for such sightings; such as aircraft, meteors, etc. However, after a thorough investigation, it was determined that the young couple were the victims of an illusion. A nearby observatory with an 85 foot radio telescope is clearly visible from the road on which the couple were travelling at the time of the observation. The radar dish was looking at the southwest horizon, and was illuminated by the light of a low full moon and a floodlight. A nearby radio tower with its red aircraft warning light is directly in line with the radio telescope as viewed from the couple's vantage point during the sighting. The dish was being rotated from the southwest horizon to the zenith so that an observer on the ground would see less and less of its surface and more or its birdcage-like understructure. When the dish reached the zenith, the floodlights were turned off, accounting for the illusion that the dish had zoomed off into the sky. All the facts described by the witnesses were accurate except that they mistook the turning off of lights, and the changes in relative bearing due to the distance of the telescope from the road, for motion.

On March 7, 1960 at approximately 8:00 PM, people from the Lake Erie area to Miami, Florida reported a bright object which flashed across the heavens. Immediately speculation was that it was a meteor. However, an investigation was begun and on March 15, 1960 the National Space Surveillance Control Center at Bedford, Massachusetts announced that the satellite carrier of Discoverer VIII had re-entered the earth's atmosphere and decomposed on March 7 at 8 PM. Sometimes it takes time but generally these UFO sightings are finally solved.

And so it goes. Sightings come in from all over the world from various types of individuals, with the great majority of the sightings logically explained after objective investigation. Certainly the experience gained over the past 13 years points up to the fact that flying saucers are not space craft from other worlds but, rather, represent conventional objects or aerial phenomena seen under confusing conditions.

4
THE PSYCHOLOGY OF SAUCERS

Dr. Carl G. Jung, noted psychiatrist and analyst wrote a book in 1959 on flying saucers, stressing the psychological reasons for the sightings. Dr. Jung states that "to believe that UFOs are real suits the general opinion, whereas disbelief is to be discouraged. This creates the impression that there is a tendency over the world to believe in saucers and to want them to be real, unconsciously helped along by a press that otherwise has no sympathy with the phenomena. This remarkable fact in itself surely merits the psychologist's interest. Why should it be more desirable for saucers to exist than not?"

Dr. Jung points out that in the middle ages religious or mythological interpretations would have been given to such signs in the heavens. Now, because of our technological advances and the possibility that man will soon conquer space, the projects or visions of man are interpreted as spaceships or flying saucers, rather than manifestations of divine intervention. Air Force experience tends to support Dr. Jung's thesis. Many letters are received by various governmental agencies on the subject of flying saucers, spaceships, and space travel. Some merely stress the technological possibilities of the aerospace age, while others feature a definite religious application or fanaticism to our space era, flying saucers and the existence of intelligent beings in other words. Here are examples of a few such letters received, with personal data and other identifying remarks edited out for obvious reasons.

November 7, 1957

TO THE AIR FORCE CHIEF OF STAFF

Respectfully refer you to previous correspondence with Major Tacker, USAF Information office. We have scheduled contact with Venusians and Martians leading us to request you take initiative with us and obtain knowledge of flying saucers plans before Communist activities claim them frightening our people who must know that space people are unarmed, friendly and will land shortly, helping us avoid war. This gives you tremendous propaganda advantage.

The Psychology of Saucers

This message was received in the Pentagon from a group of people in the southeast section of the United States. They claimed direct communication with Venusians and Martians and they furnished the names of these individuals they corresponded with from other planets. The Air Force actually checked out the claims of this particular group only to find they conducted all of their communication through a spiritualist. It was the old story of the psychic trance and assumption of the mind and character of another person, this time the other person was from another planet. In addition, this group claimed contact with many prominent deceased persons such as Dr. Albert Einstein. The Air Force simply told them they were not interested in this form of communication.

December 5, 1955

"My dear ——:

For some months I have wanted to write you but time would not permit. I am taking time today for, as an American, I feel I have the right (no, rather duty) to express how so many feel about Flying Saucers and the Government's very obvious attempt to withhold the truth from them.

For five years, but more especially the past two, I have bought every book on UFOs available and have clipped all articles from newspapers and magazines pertinent to the subject. I have written the authors of some of the books and have had a personal discussion with one.

Yesterday I purchased Major Keyhoe's book "The Flying Saucer Conspiracy," and when I finished it this morning I knew for sure I had to write. This is not something one wires his Congressman about. Since you, Sir, have been quoted in newspapers on this subject, I decided to write you.

To get to my point, I am "fed up" and disgusted with the mendacity the Air Force and other government agencies are displaying about the UFOs and I know I speak for many others who never will get around to writing you. America is a land of freedom and freedom of the press and speech are two of the basic liberties comprising that freedom. But what do you suppose we Americans feel toward our leaders when we know we are being deceived, that the press is only given hoax stories on saucers to "blow up" and the real ones are "watered down" and put on Page 39 at the bottom of the want ad column. Also, our finest airlines pilots and pilots in the Service are repeatedly made to look ridiculous because the reports or saucer sightings are explained away as mirages, sun dogs, weather balloons, etc. The American public (at least the majority with any comomn sense and intelligence) won't swallow it at all. We want the truth released so that everyone may know that intelligent beings from other planets and perhaps even other solar systems have come in spaceships and "saucers" and are observing us and watching our progress. Of course, it's pleasant to think that all these beings are friendly and mean no harm but that is unlikely since so many "worlds" may be involved. If we are told some are unfriendly I'm sure we could "take" that just as easily as the threat of atomic or hydrogen annihilation, if not more easily.

The Youth of this country wants the truth, too. I have three children and they find articles for me and read all the books they can and we talk as a family about the saucers all the time.

Fear of the unknown is worse by far than fear of the known. When we go to a hospital or a doctor for treatment we want to know "Will it hurt?" and if the answer is "yes," well we steel ourselves to face that what's ahead.

I feel as do millions of others, that knowing the truth would be the only way we could prepare for the day our Saucer friends (or enemies) land. The encounters that have taken place between earth men and Saucer men have been only friendly, as far as I know. Other incidents such as disappearing planes and crashes may mean otherwise but still we must be told the truth. It's a crime that just as we are on the verge of going out and beyond our little planet and our minds and eyes and hopes are turned heavenward, that the most glorious discoveries of all time are being kept from us. Advancement entails facing the unpleasant as well as the thrills. We cannot advance one step further if our government muzzles our radio commentators and gives our press misleading statements. It's an insult to the American people and it's hurting our opinion of our leaders.

I know the other planets are inhabited, I know the moon is too, and I know beings from those worlds have come here. I like the idea. Only those with awfully oversized egos will dislike the thought. They like themselves and their little private worlds too much to want to find out they are only a speck in the Universe. Christ said, "In my Father's house are many mansions" and so the truth seems further proof of God—not a blow to faith at all.

Every day I wake up hoping that today may be it—that the Government will tell us all and let us face the wonders and shocks and the risks involved.

I'm only one person writing and perhaps I have my nerve but this I believe is deep and I trust you will write me at your earliest convenience.

The year 1956 brings Mars close to us again and many more Saucers will be seen. Please stop trying to tell us we don't see them for we do or that they don't exist, for we know better. Be honest with us and we will be prouder than ever to be Americans.

Sincerely yours,

This is a very interesting letter. Fear of the unknown is stressed in the letter and a religious connotation is given to the existence of intelligent beings on other planets. The answer to this letter stated there was no evidence to date to prove the existence of life on other plants or the existence of space travel in reverse.

June 23, 1958

Dear Major Tacker:

I will try to be brief and to the point. There has been much controversy over the existence of several Air Force documents on the subject of unidentified flying objects. These documents are:

1. A 1947 ATIC document concluding that UFOs were real.
2. A 1948 ATIC document concluding that the UFOs were spaceships.
3. An Air Force document analyzing UFO maneuvers concluding the same.
4. A 1953 report conducted by the CIA and made by a panel of scientists concluding that the Air Force quadruple its project and release all their information to the public.

This may seem like a rehash of old material, but I would like to know if these documents ever did exist, and if so, if they are available to the public. Certain researchers in the UFO field claim they do exist yet several Air Force statements have claimed the opposite. This is quite confusing and any light you could shed on the matter would be appreciated.

I request a definite answer on this topic as soon as possible. Thank you for your cooperation.

Sincerely,

Here are the four main charges against the Air Force continually stressed by avid saucer believers. This individual was told the first three documents he mentioned are non-existent and that the forth document did exist but did not contain the recommendations he quoted in his letter. This report simply said the American public should be informed on this subject and the aura of mystery surrounding UFO sightings should be minimized.

January 29, 1959

Dear Sir:

I think the Air Force should have a flying saucer with an atomic generator in it and also one with a solar generator, or both. Doesn't it seem reasonable though? I thought I would let you have my idea.

Sincerely,

Evidently this person just felt the urge to write and suggest that the Air Force develop a flying saucer.

February 25, 1959

Dear Sirs:

I would like to tell of a strange flash of light that shone here tonight for about two seconds. It was about 8:15 PM when the light showed itself. About 15 or 20 minutes later a light flashed on and off on a sign. It was not a flashlight. If you know or have some information on such light I would like to have it please. Thank you.

P.S. Please answer. We are awfully worried about it.

This person is worried but obviously there is not enough information to allow for analysis and a valid conclusion. A copy of the Air Force questionnaire and a letter asking for additional information was sent this individual. No further information was received. This is characterstic of a great number of the letters received.

January 22, 1959

Officer-in-Charge
Security
Air Force

Dear Sir:

I have reason to believe you have given orders to quiet some UFO investigators and researchers. I would like to believe you are doing this for the good of the country. Why do you continue to try to make we researchers believe we are crazy? You know as well as I that there are UFOs flying over our country now. You also know that they are not a passing fancy. What are you afraid to admit?

What has Mars got to do with them? What do you think of Shaver? What do you think of his story? If there is any information you can give out please think of my group. There is something you can tell us without a bunch of rumors going wild and a lot of wild talk about the Air Force. You are only making all these UFO groups mad by being so critical of them. When an American gets mad he fights back, you should know that. If you don't start giving some answers these groups are going to keep digging until they come up with something. They might come up with something you want to keep quiet. Better you should let them know what they can know that to let them come up with something they shouldn't. Every time you shut one up you will find two to take his place. These UFO groups and UFO Journals are geting pretty closely knit and the more you fight them the closer they stick together. They do pass information along to each other as soon as they find it out and it won't take long until they have all the answers. What is in these UFOs the Government is so afraid of? What was in the ones you have? I would appreciate any information you can give me.

Yours truly,

The old charge of the Air Force withholding information is repeated. We simply sent this individual the latest press release "which plainly states the Air Force position." What else could be done? These people don't even recognize the fact that the correct scientific approach to any theory is for the advocate, individual or group to prove they are right. Not for the rest of the world to prove it isn't so. The Air Force has continually asked all these groups for their evidence and information proving that flying saucers are spaceships and that space travel by intelligent beings from other planets is an accomplished feat. To date the Air Force has received no evidence or information to substantiate these claims.

March 14, 1959

Members of our Air Force

Gentlemen:

Before I begin this letter I want to make perfectly clear that no one has more respect than I for our Air Force and its men, many of whom are sincerely dedicated to what they say is the best way to defend the country we all love. It is because of that love I write.

Also, I want it to be equally clear that I am not a sincere, if misguided, "crackpot."

You know that it is conceivable that I could have reason to believe that you have denied valuable information, concerning Unidentified Flying Objects, from the American public; even from those to whom some of them have appeared. You have done this to prevent a panic similar to the one caused by Orson Wells' drama; also because you wish to study the intrinsities of and concerning many findings . . . there are also men in your organization who are merely opposed to exposure because of simple, personal obstinacy, and this is unfortunate in a public service which holds so much trust and esteem. It is because of that trust and esteem I ask you to think of what will happen to it when the truth is told . . . and, gentlemen, even you know that one day it should and shall have to be told to all who then will trust and esteem you no longer.

Would you have your uniform, glorified by the blood of men who died for their country in it, fall into derision . . . ! Yet, how can you prevent this from happening when the public discovers the enormity of your deception? It will be of little value to protest then that you withheld this information "for their own good."

If the public was disillusioned and enraged at the once revered "mighty marines" because of the incidents at Paris Island, what do you imagine will be their feelings when they are shown that you have hidden knowledge of an entire other world from them? When they do move on into this new space era, what shall they say of the men who tried to protect them by keeping them in the old? Will any of the great things you have done for them in the past be remembered then . . . when you could have helped them to realize that they must all hang together or surely all hang separately . . . under an interplanetary society which is watching . . . and waiting. In other words, the simple truth is (a) The only way to prevent war over the Berlin Crisis, or any other crisis, is to force the countries concerned to hang together over an issue—any issue; (b) The things that bind people for the quickest, if for the shortest period of time, is fear; (c) If you tell the countries of the world that you have and can show them visiting this Earth in adequate space-vehicles, these countries will be bound together by fear as quickly as it takes for Ike to swing his golf club, Macmillan to pick up his umbrella, or for Khrushchev to become intoxicated.

Don't you see that you err by preventing this very binding fear from occurring?

Tell them now, before they hear it from another source and start turning to you in surprise and asking questions. How will you answer them?

You may or may not believe that I am telling the truth when I say that very soon the world shall be told that others on Earth know that other worlds in space exist. Perhaps, I shouldn't care whether or not you believe me . . . anyway, as I have said, much of the truth shall be shown soon and I shall be satisfied. I only write now to save our Air organization from the shame and disgust it shall soon face from the Earth! I, like every citizen in America, love my country and her protective services which have sprung from patriotism, a sense of duty, and all that is noble in our blessed Land. I love my country, but because I do I must chastise those who serve her, under God.

For over ten long years now we citizens have heard of U.F.O.'s and their possible interplanetary origin; we are as prepared for the good news now as we will ever be . . .

I hope you'll make the wise move, gentlemen, to save the good name of a fine organization.

You have until on or about April 20; then Russia will make a series of statements on outer-space which shall soon be followed from these and other sources, by reports to the whole world consisting of all you have tried to hide since Major Keyhoe's 1953 book came out.

Tell the world of these proofs you have of the existence of the good and kindly people from outer space—before a war shall come and bring the knowledge that could have prevented it . . . and bringing on you the curse of any men still alive after a nuclear war, still alive enough to include in a history of our worlds these accusations:

"They could have saved us but they were too blind to see wherein . . ."

I write to you as a brother—and showing the love and wisdom of the same God who rules you, me, and those from outer space.

In the holy bible, Jesus said: "In my Father's House there are many kingdoms." Help us by telling first the knowledge you have about men from another kingdom in God's House. . . .

> Because all of you are not kind,
> I must simply remain,
> In Brotherly Love . . .
> A citizen.

This letter has all the ingredients. The charges of withholding information, the fear of the unknown, desire for peace on earth and an apparent way to solve the problem, are all included. It just goes to prove we do live in a tense world and wishful thinking does convince some people that there is an easy way out. The letter offers no proof of the existence of spaceships, space travel and space people. The author strongly believes these things as a pure act of faith.

August 5, 1959

Dear Sir:

Have you tried using a lie detector—truth serum or hypnotism on these people who claim to have seen flying saucers?

A brief note from a disbeliever? Maybe not. It could be a believer seeking more tangible reasons than just his or her reliance on pure faith.

May 20, 1959

Dear Sir:

I have been doing a considerable amount of study on the subject of Unidentified Flying Objects. Your help would be appreciated very much if you could supply me with any current information. During your research work, have you ever detected any form of consistent radio waves either from the planet Mars or the planet Venus? In my studies I have also found that during the past year there has been more and more reliable opinions given that our concept of conditions on the moon may be wrong to a very great degree and that there is a possibility that much of our aerial phenomena may be originating from the moon. Does the research department lend any credence to the books written by George Adamiski?

Is it possible that the thousands of sightings around the world are all figments of the imagination?

Do you have any literature explaining your concept of these phenomena?

Is it possible that the recent crashes of our East Coast jets are due to accidents attributed to accidental contact with our own new type of coastal defenses? Have the failure of some of our missiles been deliberate to test these defenses?

I would appreciate your answer to any or all of these questions if at all possible, providing they do not violate any ruling as far as national security is concerned.

Sincerely,

This letter really covers a wide range of questions. The writer was sent the latest Department of Defense press release on the subject of unidentified flying objects. Obviously there is no answer to some of his questions and the tone of the letter indicates a tense individual afraid of what the future may hold.

March 28, 1956

Office of Public Information
Washington 25, D. C.

Dear ———:

You may recall that I wrote you in 1953 regarding light energy and UFOs. I first want to sincerely thank you for your considerate letter.

I am sincerely hoping that this letter reaches you because I feel suggestions I have to make are of utmost importance for the good of the Air Force, and the United States as a whole.

Recently I have read a most interseting and inspiring book. "The Flying Saucer Conspiracy," Author: Major Donald E. Keyhoe, U. S. Marine Corps, Ret.

I know that you are familiar with the book because Major Keyhoe mentions you in it.

What I want to suggest is this censorship business. I realize that it is all done in good faith because the Air Force thinks by revealing the facts it will cause a general panic.

However, I feel they are completely wrong here. The Americans are not

The Psychology of Saucers

cowards and they never have been. I agree with Major Keyhoe that the UFOs are from outer space. A man who is good enough to be a Major in the United States Marines certainly is not going to make a fool of himself unless he is certain.

On the other hand the fact that an H-bomb can possibly come sailing out of the blue from an enemy country of this world does not panic the people. I am sure that a warlike people from outer space will scare them even less.

The only time I do approve of censorship is to keep Russia from obtaining a secret and advanced weapons we may have. Even at this I don't think it will be harmful to disclose the fact that a nation has an advanced weapon or aircraft as long as the blueprints are kept a secret and the machine itself kept from prying eyes.

It is logical to reason that no nation on earth yet has such advanced space craft as the UFOs because if any enemy had them they would have used them against us long ago, and if we had such a craft all propeller and jets would immediately become obsolete and out of date.

Also, the UFOs whatever they are, are not hostile. If so they would have wiped us out long ago.

True, mysterious and bizarre cases have happened, like the complete disappearances of the planes in Major Keyhoe's book—the F-89, the six Navy Torpedo bombers and the Martin bomber.

Fantastic as it is these may have been taken aboard a space craft for examination rather than being destroyed and they may be released in the future alive and intact.

I feel, too, from Major Keyhoe's book that the Air Force is holding back a tremendous lot. They may even have a captured Space Craft under examination, although I don't think this is so likely because the UFOs have always been able to run away from the fastest jet.

Another suggestion I have is this, and believe me it is sincerely meant for the good.

Suppose for example things were the other way around. That is, suppose for example we were at the controls of a spaceship on Mars or any other planet, and inferior aircraft started chasing you. Would you not do about the same things as the UFOs do here? You may even shoot them down because they appear hostile.

It seems to be a mistake to chase these craft. What's the use, you can't catch them anyhow. My suggestion is this, and it is only a suggestion—I am not trying to tell anyone what to do. I suggest that on the next sighting some effort should be made to communicate with these strange craft. Instead of chasing them in fast jets I suggest that an open cockpit propeller-driven trainer go up, and wave flags at the strange craft inviting them to land. This may cause suspicion but on the other hand they may see the point. I think it is worth trying. Also on the other hand I think the pilots involved would be taking no more risk than the pilots in the jets did. The UFOs I think are definitely not hostile. If they were they could blast a jet out of the sky just as quickly as they could the propeller-driven trainer.

I feel if contact can be made with these people they can teach us a lot and certainly to our advantage. That's if contact is not already made and the Air Force is keeping it secret.

I feel it is really more harmful to withhold truth than to let it out unless this is done for the sole purpose of keeping Russia from getting something we have which I don't think is so likely now.

I sincerely hope that this letter is not annoying to you. It is sincerely meant to be helpful. It does seem that you and Major Keyhoe do not agree on some points.

Here's hoping all is well, and to possibly hear from you.

One more point I realize is this. Unless the Air Force has definite facts they can't make a statement for fear of ridicule. They have to be able to prove what they say. But I now realize that as Major Keyhoe says, facts are being withheld all in good faith, but mistakenly so. In Democracy the people are entitled to know and I go along with Major Keyhoe on this.

Sincerely,

P.S. I have not seen any strange aircraft or space craft as yet.

This individual tells the Air Force right away he believes that UFOs are from outer space and he thinks the Air Force and the Department of Defense are withholding information on the subject of flying saucers. He hasn't seen any himself but again, as an act of faith, he believes in their existence.

January 13, 1959

Dear Sir:

To a friend and ally of the United States, I would like to ask the following questions, to which I would like correct answers:

The new infra-red, heat tracker, are there any good results on recent UFOs, have any startling pictures been taken of UFOs, with the new rocket-tracking, long-range cameras, which you use at rocket-launching bases, what progress have you on the closest twin to our sun Alpha centauri 4.3 L.Y. Do UFOs come from its orbit; what news have you of the mysterious satellite still revolving around the earth.

Has any new information come out of the burning road, what metal did your investigators take away. Do the USAF consider UFOs interplanetary vehicles; will you please forward me the answers if my questions do not come under the secret act.

Thank you.

The latest press release answered this letter.

(undated)

Dear Sirs:

I have been reading a lot of books on flying saucers. I'm sure you heard that report on George Amaski, the fellow who wrote "Flying Saucers Have Landed." I would like information on that fellow who claims he was from Venus. I have studied a lot about the planets. There are people on the other planets. Venus has 7 oceans. HO2 is what they are made out of. The water on Mars, Venus are more oily-like. If you wish to know how these space craft operate they use electro-magnetic power. Remember, there's an equal reaction to every other reaction. You say your scientists are wrong about the sun, you say the further away from the sun the colder it gets. You are wrong.

Light is not force. It's just carrier of heat and energy. Sun rays carry just as much heat to other planets. Some planets might have their own heat. How, under great pressure of air? A hot core of lava. How would I know if there were seven oceans on Venus? Simple. In the book Inside the Spaceships, George Amanski tells how he talks with men from another planet. Give him any tests you want to find out the truth.

Please send me the man's address who said he was from the planet Venus. You and other men like you say that there could be no life on the other planets. You are as backward as the people in the 15th Century. They used to say the world was square. If you want to win the next war you better start working on how electro-magnetic power operates. I'm no crackpot. Thrust is not the only way to get spaceships out of earth and to other planets. If you and your scientists use electro-magnetic power, you will be able to send any amount of weight. No longer would you have to worry about weight in one of these spaceships or flying saucers. You would be able to hold thousands of pounds or even millions. You will be able to build spaceships to hold thousands of people. You will be able to see planets a hundred times closer by curving the light rays. Remember you will be creating your own gravity. The heat problem is simple.

Please send back a letter. There's a force outside each electro-magetic power spaceship.

Sincerely yours,

This letter is rather abusive and incoherent but the writer is a believer and he doesn't think much of his Air Force.

June 2, 1956

Department of the Air Force
Office of Public Information
Washington, D. C.

Dear Sir:

It is my humble opinion that, if there is the slightest bit of indication that Earth has been visited by beings from other planets, then the subject matter of such indication should be handled with the utmost care in its presentation to the world public. There should be unjaded truth about it all the way.

The recent motion picture documentary "The Truth About Unidentified Objects," by Greene-Rouse, shows in some of its footage the picture of Harry Truman on the cover of the April 7, 1952, edition of "Life" magazine, with a headline: "There is a case for interplanetary flying saucers."

This is an authentic duplication of the magazine's appearance, save for a little fallacy that should be of interest to skeptic and believer alike. The cover picture of that particular issue was not that of Harry Truman. It was that of Marilyn Monroe.

The motive for the switch might have been:
1. A desire to impress the audience that the "Life" saucer scoop occurred back in 1952 when Harry Truman was President.
2. A fear that Miss Monroe's sex appeal would be too terrestial, a defamation of the subject matter of unidentified flying objects.

Whichever it was, it probably never occurred to the movie producers that some egghead like me would run to the nearest library to check on UFOs' authenticity.

If the UFO beings do not like sex appeal, then as far as I am concerned they can stay on Mars.

Sincerely yours,

No comment!

(undated)

Dear Sir:

I accuse the press of this world of suppressing the truth concerning the danger of Atomic Explosions. Why have they not published the truth concerning the danger of radiation, causing chain reaction. In many cases, air crashes are due to this radiation, mysterious explosions in ships, fires breaking out, on land and sea, which cannot be accounted for by ordinary means.

Doctors and scientists try to explain it away by calling it a phobia, to ease their conscience. Bluffing the masses, fooling the people, denying them the truth, which has been given to them. Hypnotizing the people, pandering to man's ignorance of the true position.

The press and editors have had the truth put before them, why have they not published these warnings. Those responsible for this, will have to account for this suppression when passing into the next sphere of existence.

You had the power of the Press to tell the truth, you have failed in your duty of publishing these great truths.

THE MASTER AETHERIUS SPEAKS TO EARTH FROM VENUS

"It is for this reason that our major approaches have been made to those who may appear to all intents and purposes to be *ordinary individuals*.

"We have been asked several times, why do we not channel our approach through scientific organizations. My answer to that must be that the majority —not all—but the majority of Terrestrial scientists have minds likened to a cup filled with liquid which cannot hold anything more. Any other knowledge offered to the minds of certain scientists upon Earth, would be purely and completely representative of a waste of mental energy.

"When the cup is full, more liquid is a superfluity for it runneth over.

"It is for this reason that we are aiming an approach to the masses, in dozens of different ways—some apparent, some not so apparent. It is because the ordinary decent, right-thinking, World Citizens—note that please— WORLD CITIZENS—are the people who will have to act as one great whole, in order to put right this terrible chaos which now reigns upon your Earth.

"These are the people who suffer by this chaos. In some countries they suffer starvation and disease. In other countries they suffer an ignorance which has been specially planned—the foulest move of all this! They suffer dictatorship, in other countries. They are conditioned and have been, throughout the centuries. It is the ordinary man who is the sufferer when war comes. The ordinary man does not gain anything, one way or the other. He is the loser in war—and also the person conditioned in Peace.

"It is for this reason that we are making our main approach to the ordinary individual because the Lords of Karma have stated that beyond all doubt, that it is this sufferer who will eventually break the bond which has bound him as a slave to dreadful orthodox conditioning for many lives.

"The revolution will be a mental one. It will be a Renaissance. It will be a transmutation of basic thought into Spiritual action. Note that—A TRANSMUTATION OF BASIC THOUGHT INTO SPIRITUAL ACTION. That is what it will be! That is why our approach has been directed to you, who are the backbone of your Earth. To you, who are on the mental Realms and to you who are the backbone of your Spiritual Realms.

"We make our main approach to you and in doing so remind you again, of your great responsibility to your Earth and to those who are deaf and blind—in the higher sense—upon your Earth.

"These need your prayers and your Healing, for they are the great black ones. Their magic has been specially designed throughout the centuries to make powerful castles for themselves. They have done this. Now they manipulate Governments, whole countries and these countries obey like so many puppets at the bottom of the strings in a marionette show!

"This dark group will be taken from the centre of your Earth but dear friends, please, PLEASE, do not let their evil schemes bear fruit before they are taken from it. By that I mean, that the great conspirators are in your midst. They have been there for centuries. You have danced to their tune! The whole propaganda organization throughout all Earth, which produces either an uneasy Peace or a war or a cold war or fear or famine, is the tool used by these dark few among you. They are very powerful for they manipulate the monetary wealth of Earth. This gives them almost unlimited power upon the surface of Earth.

"But note this, Earth! The time of their trial is nigh! The time when a great beam of Understanding and Transmuting Light, which will be thrown deep into the heart of this foul, black, cancerous growth within your Earth, is shortly due to come.

"The dawn, my friends, will soon break.

"Be ready for it—when it does so!

"This is another main reason why we have directed our major appeal through Primary Terrestrial Mental Channel and why this major appeal has over-ruled all other contacts, all other appeals. Why we have directed it towards the decent World Citizen, in the hope that he will allow his decency to rise uppermost within his heart and will sink petty difference, burn up so-called individuality in the fire of total co-operation and be prepared for the coming of the Great Transmuting Light.

"I said a few moments ago, that you must not allow the plans of the few dark ones, to bear fruit. By this I mean that, even at this very moment they are trying to stir up strife within this Earth of yours. To cause a war between countries and hate to exist between white men, yellow men and coloured men. They are trying to do this!

"They will succeed. They will succeed in bringing about a wholesale war to the face of your Globe. They will succeed in bringing about an atomic war to the surface of your Globe, which will destroy and horribly mutate your children for a hundred generations—UNLESS—you stop them! Unless you stop them by non-co-operation with their foul plans."

WARNING FROM SPACE
TREAD CAREFULLY YE MEN!

"Earth today, stands on the very brink of the first attempt to probe the mysteries of outer space.

"The men of science just took it for granted that the people who inhabit other Planets and have used your Moon as a base for nineteen million years, would welcome them with open arms.

"Either that—or they cared not whether the users of the Moon wish their trespass!

"Although the bases of the Moon have been used primarily so that Earth

The Psychology of Saucers

should be protected from her own wrong-doing, those who asked us not, knew of our positions upon Luna. Those who did not know of our positions were ignorant only beacuse they chose to hide their greying heads beneath the sands of their own dogmatic ignorance.

"According to our agents, the United States of America intend—quite shortly—to launch a rocket towards the Moon. This is to be followed soon afterwards by an attempt which will be made by Russia. In neither case has our sanction been sought.

"We forgive you this—for you are children. But I would remind you, Sirs, that when you place a vehicle in that position in space you call—free-fall, responsibility towards all life and towards the Cosmic Whole, is increased three million times.

"To ignore this warning will not alter the fact! You have been told!

"If you try, at any time, to bombard the Moon with an explosive weapon you must reap the inevitable—listen to that, Earth!—inevitable consequences.

"The United States of America profess belief in the existence—and even Deity of Jesus. Then Americans, may I remind you of what that Being states.

"He said: 'As a man soweth, so shall he reap!'

"THIS IS LAW!

"If you, or any other country upon Earth make any attempt to bombard the Moon or any other inhabited Planet or Satellite, with explosive weapons, you will reap the consequences almost at once!

"You may turn what deaf ear you wish to this. The fact is; and forever will remain so.

"I would say to American scientists, immediately you venture outside the gravitational field of your own Planet, Earth, your behavior will be most carefully watched. Not by people from Mars or Venus but by greater people than these—by the Supreme Lords of Karma. By the self-same Mighty Galactic Beings who guarantee that the Law of Action and Reaction is perfect!

"If you go into Space you must behave yourselves. If you do not do so, Divine Justice is liable to strike like lightning.

"Those of you who can think for yourselves, must see the logic in this. Must be eternally grateful for this kindly advice. Because, Men of Science, this is advice—and not warning.

"If you murder another Earth man, Your Karma will demand its balance —but it may do this some time after. If you murder any other Intelligence from a more highly advanced Planet than your own—your Karma must find its balance. It will do this at once.

"I would advise you before you attempt to throw an atomic missile at your Moon, to think well upon the teachings of the Master you say you follow. He stated that, 'He who lives by the sword, shall die by the sword.'

"Throw a bomb in to the Serene face of the Moon, Earth, and you will die! In the self-same way Karma will extract balance. This is the Law, whether you like it or not! Whether you believe it or not!

"In this respect now, ignorance can no longer be your glib excuse.

"FOR—I HAVE SPOKEN!

"This Transmission was an emergency Transmission to the Space scientists of Earth.

"My friends—if you ever do circumnavigate the Moon, please realize that it is the people on the other Planets who have allowed you to do this. In return, you must behave yourselves—in a very definite, very controlled way, or else you must be prepared to take the Divine consequences."

This is a lengthy letter with a quoted message from Venus warning the people on earth of their responsibilities in space. This was a circular letter with no return address and could not be answered although it is very doubtful if it would have been answered had there been a return address.

There is a great fear of another world war shown in this letter and note the reasoning used to explain why their appeal is to all ordinary individuals instead of to scientists and governmental representatives. Also it is again interesting to note that we earthlings are but children compared to the superior intelligent beings of other planets.

Many of the people interested in the subject of unidentified flying objects, popularly termed "flying saucers," of spaceships and space travel have identified themselves with formal clubs or groups whose mission or aim is to conduct research in this area.

In this respect there are over one hundred large organizations of this type in existence today. Most of these organizations have annual dues and publish periodical papers or newspapers on the subject of flying saucers.

On the one hand, some of these groups identify themselves as "ufologists" who say they are scientific or semi-scientific groups dedicated to objective research in this scientific field. While on the other hand, there are the so-called "saucerian" or "contact groups" who claim they have talked with people from outer space and ridden in their spaceships. Strange as it may seem these two groups bitterly attack each other even though they share a common belief that spaceships and space people from the other planets exist.

As mentioned before, a critical examination of sightings reported to the United States Air Force has revealed that a good percentage of the reports were submitted by serious people, mystified by what they had seen and motivated by a patriotic responsibility. The record also indicates that there are individuals associated with UFO organizations who are interested in

The Psychology of Saucers

unidentified flying objects or flying saucers because of the dollar profit involved in the sale of newsletters, books, motion pictures and still photography, purporting to prove that life exists on the other planets and that the space people are watching us with a great deal of interest. In most cases the writers indicate that the people from these other worlds are much more intelligent than earth people, intend us no harm and, in fact, may step in at exactly the right moment and save us from ourselves. This, of course, is opportune for these writers and is a reflection of the tense world we live in under the ominous threat of a nuclear war which they say will destroy our civilization.

It is important to state once again that the United States Air Force does not deny the possibility that life could exist on other planes or in other solar sysems, or that conditions for life as we know it on earth could not exist somewhere out in space.

The Air Force simply states it must be objective in its pursuit of the truth in investigating, analyzing, and evaluating each of the sightings reported to its various activities and to date there is no evidence to substantiate these claims of interplanetary or interstellar travel.

The Air Force emphasizes the belief that if more immediate detailed objective observational data could have been obtained on the unexplained flying saucer sightings in its files, these too would have been satisfactorily explained as conventional objects or some form of aerial phenomena.

The senseless and vicious attack against the Air Force charging its organization with withholding vital information from the general public is a baseless charge, is absolutely untrue and is generally used to excite interest in a sensational news article, magazine article or book.

From time to time various members of the Congress and Congressional Committees have evidenced an interest in the subject of unidentified flying objects. Their interest is, of course, engendered by letters and telegrams from constituents who evidently believe in the existence of spaceships and are convinced that the unidentified flying objects seen in our atmosphere are spaceships and constitute a threat to the security of the United States or they are convinced that the United States Air Force and the United States Government are withholding vital information from them as citizens. Many of the letters coming in to the Congress have been prompted by planned campaigns on the part of UFO or Flying Saucer Clubs and groups demanding public hearings on this controversial subject.

A typical letter used by these organizations follows. This letter had an attachment stating that spaceships are fact, not fiction, and referring to a

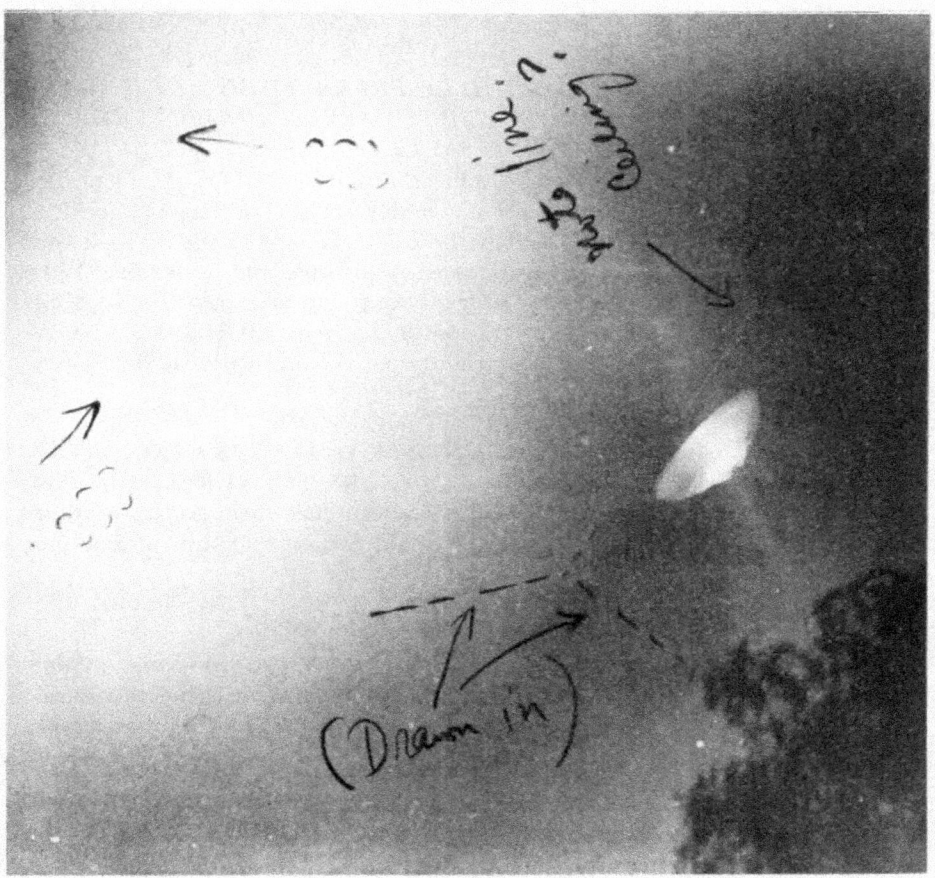

This picture is a hoax submitted to the Air Force for evaluation. Upon close examination it was found that the individual had taken a time exposure of the full moon, covered the lens, then took the camera inside and took a picture of a light fixture; covered the lens again, and took the camera outside again. An unsuccessful attempt was made to set the camera in the exact spot as before. The moon's track on the negative was not linear. The edge of the ceiling is outlined above the fixture.

so-called actual contact at Kearney, Nebraska on November 5, 1957 and Elm Creek, Nebraska on February 5, 1958. The Elm Creek incident was not reported to the United States Air Force. The Kearney, Nebraska incident was reported to the Air Force and was thoroughly investigated by representatives of the Air Force. The final conclusion in this case was that the incident was an outright hoax. This was also confirmed by local law enforcement authorities in Kearney.

Here is the letter which was sent to all saucer enthusiasts by UFO clubs and groups along with a crude drawing of a saucer.

"WHAT YOU CAN DO TO HELP BREAK THROUGH THE SECRECY ON THE UNIDENTIFIED FLYING OBJECTS OR FLYING SAUCERS

"It is known that Radar Reports and other data on UFO's (Unidentified Flying Objects) or Flying Saucers is under the wraps of secrecy in the Military. All information is collected but the public is kept in the dark on this vital subject. What information is released is distorted and misleading. Why should this be so, ask the alert, intelligent American who pays the TAB for all of this. There is a rising demand for the Citizens RIGHT TO KNOW about all of the hidden facts and rightly so. A constitutional democracy cannot endure without a informed electorate.

THIS IS WHAT YOU CAN DO

"1. Write intelligent letters to your STATE SENATORS & CONGRESSMEN—also your UNITED STATES SENATORS & CONGRESSMEN. Ask for a free flow of information on this subject. Ask them to act as your elected Representatives in this matter to make certain that this information reaches the public.

"2. Tell them about local interest, local sightings on UFO's, photos, Clubs and activities.

"3. Ask that a civilian board be appointed to collect this information and release it to the public. Let the people evaluate these reports and this information. It belongs to the people.

"4. Ask for an investigation by:

> The Armed Services Committee—Sen. Richard B. Russell, Ga., Chairman
>
> Senate Committee on Gov't Operations, Sen. John L. McClellan, Ark., Chairman
>
> House Committee on Gov't Operations, Congressman John E. Moss, Chairman
>
> Gov't Appropriations, Sen. Carl Hayden, Ariz., Chairman (Senate)
>
> Gov't Appropriations, Sen. Clarence Cannon, Mo., Chairman (House)

If the right committee for the job (Interplanetary Study) does not exist, ask that immediate steps be taken to accomplish this. Major Wayne Aho Washington Saucer Intelligence, P. O. Box 815, Washington 4, D. C., has been to every Senator's office and every Congressman's office on this subject, and is continuing his work on Capitol Hill, until results are obtained. He has been told by government officials in the Senate and House that there is NO COMMITTEE IN EXISTENCE that is qualified for this specific assignment. Interplanetary Research is unique and new to our time. It is time for *action* to correct this weakness.

How else can we enter the *Age of Space?*

"5. It is time for us to face the issue squarely. The time has come for forthright answers. Let us end the doubletalk and evasion.

"6. Request a REPLY

If America is to enter the SPACE AGE with honor, let us face up to the responsibilities of the Space Age. Let us—rather than attempt to conquer space begin to UNDERSTAND Space. With this approach—we too can become SPACE TRAVELERS.

Here are two forms to address your letters:

For a Senator: Honorable Charles E. Potter
                 United States Senate, Washington 25, D. C.
For a Congressman: Honorable Gerald R. Ford, Jr.
                     House of Representatives, Washington 25, D. C.

Dear Sir—or Dear Senator Potter—Very truly yours,

Dear Sir—or Dear Mr. Ford—Very truly yours,

"This is how YOU CAN Help in bringing about many wonderful events.
                 NOW Let us hear from you.
SUPPORT YOUR LOCAL CLUBS, ACTIVITIES, PUBLICATIONS"

Through such letters UFO clubs and groups have succeeded in creating widespread interest over the past 13 years and certain Congressional Committees, such as the Senate Committee on Space and Aeronautics and the House Committee on Space and Aeronautics could not ignore this particular subject when preparing our new space bills. Therefore, these Committees requested and received comprehensive orientation briefings on this subject by Air Force teams. All Committees who have been briefed by the Air Force on this subject, based upon their interest for the preparation of space legislation or interest engendered by the demands of their constituents, indicated that they were satisfied with the Air Force program to investigate these sightings and believed the program was in competent, capable hands.

Committee members who have received these informal briefings during the past two years were associated with the Senate Committee on Space and Astronautics, the Senate Subcommittee on Government Operations, the Senate Preparedness Subcommittee of the Senate Armed Services Committee, the House Armed Services Committee and the House Committee on Science and Astronautics. In addition many individual Senators and Congressmen received either briefings on this subject or letters and press releases outlining the Air Force findings and its position on the subject of unidentified flying objects.

In all its dealings with the Congress, the United States Air Force has continuously stated that should overriding considerations develop which would necessitate public hearings on this subject, the Air Force stands ready to cooperate in every way with the Congressional Committee involved.

5
IT'S EASY TO BE FOOLED

Air Force Regulation 200-2 (See Appendix One) defines an unidentified flying object as any airborne object which by performance, aerodynamic characteristics, or unusual features does not conform to known aircraft or missiles, or which cannot be positively identified as a familiar object.

Reports of unidentified flying objects must be accurate and complete, otherwise analysis and evaluation are very difficult. In some instances a reported UFO could have intelligence value and this is just another reason for accuracy and complete reporting on the part of the observer. (See Appendices Two and Three)

Unusual weather or light conditions may make many familiar objects into unidentified flying objects. The speed of the observer's aircraft and sudden climb or descent may produce distortions of vision which cause known objects to hover, perform erratic maneuvers, or glow and scintillate during hours of darkness. Many of these flying objects can be identified as conventional aircraft observed from unusual angles; modern jet aircraft flying at great speeds and high altitudes; reflections of sunlight, moonlight, and starlight from aircraft and balloons at great heights; searchlight reflections on clouds; meteorological and upper air research balloons; meteors, comets, and stars; planets observed at certain times of the year; meteorological phenomena; cloud formations; birds, especially migratory formations; dust and haze; kites, fireworks, and flares; rockets; and condensation trails.

A meteor, a comet, a balloon, or an aircraft, under certain conditions, assumes speeds, movements and shapes which are entirely uncharacteristic of the object under normal circumstances. Aircraft at great heights can appear wingless and projectile-shaped. Objects that appear to hover or move very slowly could be balloons. Flame-tinged, or brightly-glowing objects, and those objects appearing to leave a trail of light in their wake may frequently be identified as meteors or comets. Another explainable phenomenon may be caused by the sun's illumination of vapor trails. Moving lights at night, or shiny objects in the daytime, travelling at moderately fast speeds, could be aircraft.

It has been characteristic of many reported observations of unidentified

52　　　　　　　　　　FLYING SAUCERS AND THE U. S. AIR FORCE

A rare cloud formation sometimes referred to as a lenticular cloud.

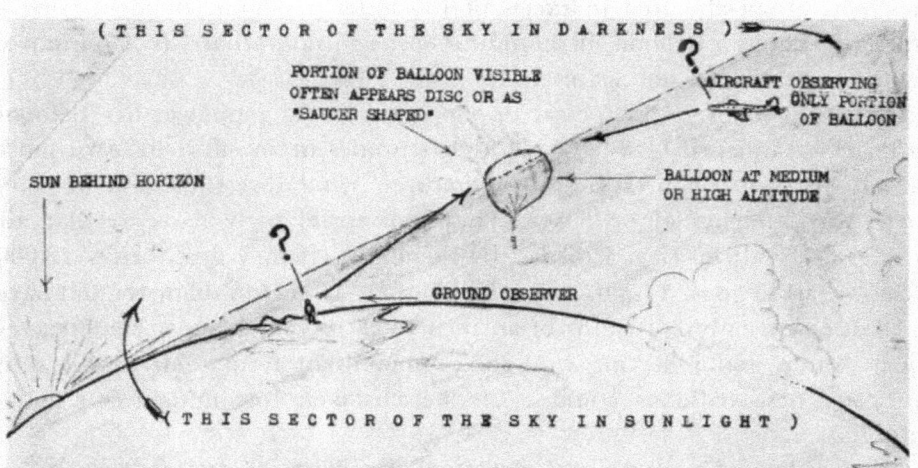

The above sketch clearly shows how unusual lighting conditions could fool a ground observer. The sketch is exaggerated to emphasize this phenomenon.

It's Easy to Be Fooled

flying objects in the past that they have indicated at least some features of modern aircraft. There have been descriptions, including rocket or jet pods, fins or rudders, windows or portholes, propellers, exhausts, etc. High speeds or modern-day aircraft lessen the possibility of detailed observation, and only certain prominent or familiar features of the flying object may stand out in the observer's memory.

Silvery, transparent, disk-like objects may be balloons. The absence of exhaust or engine noise, or any visible means of propulsion, would support such identification. Weather balloons are often released in clusters and many drift in what appears to be formation, depending on the air currents. They shimmer in reflected sunlight or moonlight, and seem to hover as they pass from one air current to another.

Upper air research balloons may attain great heights and travel great distances before they burst and fall back to earth. They may be observed, therefore, in areas far removed from any logical launching site. Research balloons are usually constructed of material with a highly reflective surface. They often approximate one hundred feet in diameter and are visible, under certain atmospheric conditions, even at extreme heights. Such balloons, seen in reflected light, may seem disk-like in shape and many appear to have an oscillating motion. They carry metallic equipment which can result in electronic contact.

An object usually is not a balloon if its speed is too fast. However, some balloons, such as those used for cosmic research, travel in the upper air currents at speeds often in excess of 100 miles per hour. In identifying a flying object as a balloon, it should be borne in mind that a balloon moves with the wind and not against it.

In the field of technological developments, new giant weather baloons are being launched to fly at very high altitudes in an effort to learn more about atmospheric pressures, temperatures, wind directions and velocity over vast stretches of open sea. They will travel high above regular air routes and will be rigged to destroy themselves if they drop below 28,000 feet, or fail to go that high. These balloons are 40 feet in diameter and have a plastic skin only 2/1000ths of an inch thick. Flying at great heights over open water, and reflecting sunlight or moonlight from their plastic skin surface, these balloons could easily be mistaken for unidentified flying objects.

The estimated azimuth and elevation of a flying object can be checked to determine the known location of astronomical bodies. Meteors may be

An experimental high altitude balloon in a strong wind assumes unusual shapes.

identified by appearance, great speed, short duration of sighting, and in instances of fireballs, color and brightness. At the time when the planet Venus is very low on the horizon, it can appear to change color, perform erratic maneuvers, or become distorted and diffused when viewed through thin clouds, haze, or alternate layers of warm and cold air. Meteors, on the other hand, do not pursue an erratic course. When the duration of observation of a flying object is extremely short, it is highly probable that the object is an astronomical sighting.

An unidentified flying object may assume various shapes. The four most common shapes reported in the past are elliptical or disk shape, aircraft shape, cigar shape, and propeller shape.

Shape is an important factor in determining the identity of a flying object. Distortion of shape, due to distance and darkness, enhances the difficulty of identification. Many of the strange shapes reported in the past would appear to be unidentifiable in terms of familiar objects, but in many instances could have been reflections from conventional objects viewed under unusual

It's Easy to Be Fooled

conditions. Light and shadow produce fantastic distortions, especially when objects are viewed at great distances and in varying degrees of gathering darkness.

This variety of shapes is an indication of individual reaction to what may have been familiar or conventional objects seen under unusual conditions, or created in the mind of the observer by his physiological limitations and psychological responses. Fatigue, unusual weather conditions, and the stress of flying at great speeds and high altitudes could induce such manifestations.

One report of an unidentified flying object stated that it was shaped like a conventional aircraft, but was luminous and surrounded by a red glow. This phenomenon could have been an actual aircraft reflecting light from some undetected source within or on the aircraft and glowing from an unusual play of moonlight or starlight on metal parts.

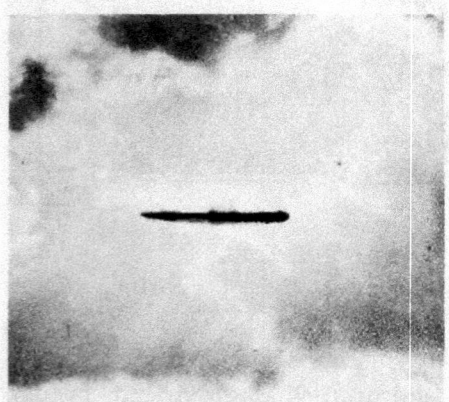

Unusual lighting conditions produced by bright sunlight and cloud cover, completely obscure the wings and tail assembly of a giant B-36 bomber, making it appear elongated or cigar shape. This is one of the most often reported shapes in flying saucer sightings.

A disk-like object, with illuminated portholes, could be a conventional aircraft distorted in shape and stripped of wings by a temperature inversion mirage effect and reflecting light through apparently dual and convergent sets of windows.

Transparent, cigar-shaped objects, illuminated from the inside and emitting an exhaust, could be jet aircraft at high altitudes where they appear wingless. The mirage effect of a temperature inversion could cause the apparent illumination and transparency.

Saucer-shaped objects, which hover and maneuver erratically, could be the planets Venus or Mars seen near the horizon at certain times of the year. When objects are viewed through haze or mist, the limitations of the human eye can produce what appears to be a hovering effect, or erratic movement.

Propeller-shaped objects could be conventional or glider aircraft, distorted in shape by mirage effects caused by a temperature inversion.

Radar sightings of flying objects frequently may be explained as ground targets reflected by a temperature-inversion layer, or as radar echoes of various objects, not all of which are visible to the human eye. Most solid objects produce radar responses which are recognizable. Moving objects, such as aircraft and birds, normally can be identified by the size of the radar blip and by the speed, altitude, and type of movement measured by the radar set. The radar operator should be able to determine whether the responses noted on his scope are real, or are caused by the weather or other phenomena. A blurred effect on the radarscope may indicate a weather target, whereas a solid target, such as an aircraft, will be sharply defined.

A radar equipped Lockheed Super Constellation. Odd shaped aircraft confuse observers.

There are many new developments in aeronautics and astronautics which contribute toward new or differently shaped flying objects. The saucer-domed radar planes, flying as picket ships on our Dew Line Extensions are an example of an unfamiliar type of aircraft which, to the uninitiated observer, might appear to fall into the unidentified flying object category.

Technical advancements in the field of supersonic rockets to gather weather data offer possibilities for radar detection of apparently unidentified flying objects. Rockets will be shot into the stratosphere at 3000-plus miles per hour. At a height of 20 miles or more, their warheads will explode and release a cloud of metal foil fragments, which will be traced by radar to learn wind velocity and direction.

It's Easy to Be Fooled

The construction and successful launching of man-made satellites into the edge of space above the earth as part of this country's aerospace program, may lead to reports of unidentified flying objects. The United States expects to launch a number of scientific data-gathering satellites over the next few years and Russia has announced intentions of exploring space to the utmost.

The satellites developed by United States scientists are highly-polished spheres to be propelled aloft by a three-stage rocket. Planned orbit is at a height of about 300 miles and is in the direction of the earth's rotation. The satellites' course will follow a path that will permit its sighting from positions in Europe, North Africa, and the Middle East. Although the satellites' size will appear small at such an extreme altitude, its reflection will be visible to the naked eye under certain weather conditions. Its terrific speed will carry it from one horizon to the other, within the view of an observer in less than 20 minutes. The Discoverer Satellites are of this type.

Many new types of aircraft are under development and some are in production at this time. Certain types will be capable of vertical take-off. The unusual configuration of these aircraft lends itself to possible confusion with unidentified flying objects, and a vertical take-off might add to the observer's failure to identify it as a known object.

Analysis thus far has failed to provide a satisfactory explanation for a number of unidentified flying objects. An understanding of some of the phenomena which may cause familiar objects to assume unfamiliar characteristics, together with an awareness of the many new technological developments which may be observed, should result in fewer sightings of this nature. Rational reporting will facilitate analysis of those sightings reported as unidentifiable.

Scientists have been exploring the mysteries of the universe for many centuries and today know a great deal about the composition of the galaxy which includes the earth among its many planets, stars, and other celestial bodies. Yet, many questions remain unanswered and the search for more knowledge in the broad field of astronomy continues. The same is true regarding the earth's atmosphere, and, although considerably more is known regarding the natural laws which govern the sea of air around the earth, there are many aspects of meteorology that are not yet fully understood.

It is not unusual for the mind to become confused by garbled messages, caused by unusual astronomical and meteorological conditions and transmitted to it by the eye. Thus, the sky has been the setting for many strange sights which were not readily understood. Many may have been the result of unusual astronomical and meteorological conditions, which cannot yet be scientifically explained. However, many types of illusions which appear to be flying objects can definitely be related to astronomical and meteorological phenomena.

Under certain atmospheric conditions, reflection and refraction processes can transform conventional aircraft, automobile lights, planets, meteors, and other identifiable phenomena into apparently supersonic flying objects of many shapes and colors. Clouds, haze, industrial smoke, weather droplets and ice particles in the atmosphere are typical ingredients which make up atmospheric lenses through which many illusions of flying objects are seen. Car lights reflected on clouds can create luminous disks which dart erratically through the sky at terrific speeds. Other light sources can produce similar illusions with appropriate variations, many of which even have specific colors provided by refraction of the light through water and ice particles in the atmosphere.

One of the most common causes for optical illusions of distorted and displaced objects is the mirage. Warm air has a lower refractive index than cold air. The air is normally warmer at the surface of the earth and progressively cooler in a fairly steady gradient through high altitudes. It is

A mirage in the high dry mountain air under brilliant sunlight looks very much like the saucers Kenneth Arnold spotted on June 24, 1947.

through such atmospheric conditions that distant objects are usually viewed and the mind becomes accustomed to the impressions conveyed to it through the eye in this normal perspective. Light rays normally travel in a concave path that intersects with the horizon. When the normal temperature distri-

It's Easy to Be Fooled

bution is upset, the light rays bend accordingly and optical phenomena result. Causes of mirages follow two basic patterns:

(1) When the surface air is exceptionally warm, the air expands and becomes less dense, causing the convex path to shorten and, under extremely hot conditions, even to become concave.

(2) Under conditions of a temperature inversion, with a layer of warm air over cold air, the path of light rays will lengthen to parallel the earth's surface at greater distances.

These abnormal conditions cause mirages and the eye will see unfamiliar or displaced and distorted images, which the mind is not immediately capable of interpreting correctly. Realistically proportioned mountains, cities and seas may be projected high into the atmosphere. On the other hand, land areas may be distorted and appear as separate images floating in the sky, giving the impression of suspended or flying objects. From an aircraft in flight, a cigar-shaped illusion of a land mass can change size drastically with changes of only a few feet in altitude of the observer's aircraft, thereby giving the illusion that the object is accelerating rapidly, travelling alternately at slow and extreme speeds, going away from the observer or coming in toward him. The same is true at night in the case of objects formed by such light sources as searchlights, glow of lights from cities, automobile headlights, and celestial bodies. A temperature inversion can reflect the image of an aircraft to another location in the sky and mirror it as two aircraft, perfectly joined, with one aircraft inverted below the other.

The common mirage, based primarily on temperature distribution, is of course, only one type of the numerous meteorological phenomena producing aerial apparitions. Others are caused by reflection and refraction of light through various atmospheric structures, such as different types of clouds, water droplets, ice and frost formations, haze and smoke. Combinations of meteorological situations, and even combinations of meteorological and astronomical conditions, can produce startling effects.

Combined refraction and dispersion of the earth's atmosphere can cause a celestial body to appear to be at a different location in space and distort its normal color as well. When the object is low on the horizon, this condition is particularly prevalent. The planet Venus, for instance, may appear as bright red on the bottom and bright blue at the top edge, thereby giving the illusion of a flying object emitting red exhaust trails. An observer flying in an aircraft may easily mistake such an apparition for a flying object. As the aircraft moves through the atmosphere at an advanced speed, its position relative to the object naturally changes and the atmospheric conditions in line of sight between the aircraft's position and the object may change as well. The object thus may assume apparent characteristics of erratic behavior and fantastic shapes and colors.

Although all the planets may resemble flying objects under certain conditions, Mercury, Venus, Mars, and Jupiter are most commonly mistaken in this sense.

At its brightest, Mercury has a stellar magnitude of —1.9; this is more brilliant than a first magnitude star. This planet can only be seen occasionally, and then only for a short period during morning or evening twilight. The reason this planet is never seen throughout the night, as are some planets, is due to its nearness to the sun. It rises and sets very close to the same time as the sun. 28 degrees is as far as it can ever get from the sun and this distance is never perpendicular to our horizon, but along a line running south of the sun, as the ecliptic (sun's apparent path around the earth) is at an angle to the horizon. Mercury's orbit is inclined seven degrees to the earth's orbit, greater than that of any planet visible to the unaided eye, but even this inclination does not take the planet very far from the ecliptic. Since Mercury has an orbit inside of the earth's, it passes between us and the sun; and since (like all the planets) it owes its light to the sun, it goes through a series of phases like the moon.

Venus, with a stellar magnitude of —4.4, is the brightest of all the planets and Mars is next. Venus, at its brightest, can be seen in daylight and can cast shadows after dark. The planet Venus is a morning star for approximately nine or ten months and then an evening star for the next nine or ten months. In 1960 Venus will be a morning star from January to June and an evening star from July through December. Nearer, larger, and brighter than Mercury, this planet is consequently much more conspicuous. Due to its greater distance from the sun, a maximum angular distance of approximately forty-six degrees, it is nearly always visible. Venus for months at a time stands out in plain view as either an evening or morning star, the brightest in the sky. The brightness of this planet is due to two factors—its proximity to the sun and earth, and its high albedo 0.59 (reflecting power). This means that Venus reflects 59% of the light which strikes it. The orbit of Venus is also inside that of the earth. It therefore passes between us and the sun, and this results in the planet having moonlike phases.

Mars is an evening star about thirteen months and a morning star for approximately thirteen months. This is the first planet outside of the earth's orbit. The albedo of this planet is approximately 0.15 and its maximum stellar magnitude of —2.7 makes its apparent brightness greater than any

It's Easy to Be Fooled

other planet except Venus. Since Mars is outside of the earth's orbit, we always see nearly all of the side on which the sun is shining. Therefore, it does not go through noticeable phases as do Venus and Mercury. When observed through a telescope it sometimes appears slightly gibbous. Mars appears reddish in color.

Jupiter is one of the brighter planets. Only Sirius, a star, and the planets Venus and Mars are brighter. The albedo of Jupiter is 0.56 and its stellar magnitude -2.3 at its brightest. Jupiter has a very large orbit about the sun as compared with earth's orbit. Therefore, Jupiter, like Mars, has no phases, but always shows us the lighted surface.

In the past, both Venus and Mars, when low on the horizon, have been observed to change color and move at fantastic speeds, when viewed through haze or mist. Venus appears low on the horizon during the spring and is unusually bright. Mars has been reported to resemble a flying object when it was low on the horizon in early summer. If one of these planets is stared at for any length of time without any balancing point of reference, it can appear to perform erratic maneuvers. Thus, the planets of brighter magnitude in our galaxy provide a constant source of illusionary flying objects.

Occasionally one of the brighter stars, such as Sirius, is reported as an unidentified flying object. On these occasions, as with the planets, the star is usually low on the horizon and the weird effects of apparent motion and color changes which are induced by atmospheric refraction and diffusion contribute heavily to the misidentification.

Halo occurs when light from the sun or the moon passes through thin upper clouds composed of ice crystals causing various circles or arcs of light to become visible. These are called solar or lunar halos, and are produced by refraction. The most common halo is a ring of 22 degrees radius around the sun or moon, both of whose angular diameters is approximately one-half a degree. Various figures may result due to the differing shapes and positions of the falling ice crystals through which the light passes. Halos are often white, but a well-developed halo is red on the inside shading off to yellow.

Comets and meteors are often mistakenly identified as flying objects, although sightings of comets are rare simply because their incidence is so low.

Comets are nebulous bodies revolving around the sun for the most part in long ellipses. Although their periods are very uncertain, some few such as Halley's Comet, which pursues unmistakable ellipses, can be expected to return. The nucleus of a comet, which is believed to be composed primarily of frozen gases, strengthens in brilliance the nearer the orbit of the comet brings it to the sun. As the comet moves closer to the sun, the ice tends to evaporate giving off gases that form a cloud around the nucleus known as a coma. Some comets become bright enough to be discerned even in daylight. Since the long tail of the typical comet is composed of matter repelled away from the sun, it may either follow or precede the head, depending on whether it is approaching or going away from the sun.

Meteors are particles entering the earth's atmosphere where they become so intensely heated they turn into incandescent gas. Theories on the origin of meteors are largely controversial; however, educated guesses range from dissipated comets to disentegrated planets. It is estimated that 24,000,000 meteors, which can be observed by the naked eye, enter the earth's atmosphere during a 24-hour period. These space particles are of various sizes, ranging from the microscopic to the rare ones weighing tons.

Bright meteors are known as fireballs or bolides. Some of these penetrate the lower parts of the atmosphere, where they explode with a noise like a distant thunder. These are rare—probably no more than a few dozen

A fireball comet streaks across the heavens at night in one of nature's unusual displays.

appear over Europe during an average year. When a meteor, of such size that it is not entirely consumed by frictional heat after it enters the atmosphere, eventually collides with the earth's surface, it is called a meteorite. It is estimated that about 2,000 of these latter enter the earth's atmosphere during an average year.

The appearance and behavior of meteors streaking through the earth's atmosphere take on various fantastic forms, depending upon their size and

It's Easy to Be Fooled

composition and the meteorological conditions through which they are viewed. A meteor with the brilliance of the Pole Star can be caused by a particle no larger than a grain of sand. A particle no bigger than a pea can become a fireball. Examination of discovered meteorites reveals that most are irregular in shape; however, many become conically shaped in their passage through the earth's dense atmosphere.

Meteors may appear as bright balls or disks with fiery tails, which could be mistaken for jet or rocket-type exhausts. It is not uncommon for meteors to appear as flaming fireballs, with colors ranging from dull red to bright green, and they may even travel in clusters, giving the appearance of flying objects in formation. Meteors may also move relatively slowly and appear to follow a path parallel to the horizon, thereby giving strength to the illusion of flying objects.

Large meteors have long paths and may cross from one horizon to the other in the view of one observer and pass far beyond. They travel in the same direction as the earth in its orbit and their speed upon entering the earth's atmosphere varies. Those meteors overtaking the earth during evening hours may travel initially as slowly as seven miles per second, while those meeting the earth's rotation head on during morning hours can be travelling more than 40 miles per second. Multiply these types of appearances and behaviors by complementary meteorological phenomena and the prospects for illusionary flying objects are considerably increased.

A phenomena which may be seen in conjunction with the sun and moon is that known as sun dogs and moon dogs respectively. This phenomena is an image of the sun produced by refraction in ice crystals and is often seen in conjunction with a well developed halo. The sun dogs most frequently experienced are on opposite sides of the sun at a distance of 22 degrees and are red next the sun. Secondary formations of this phenomena may occur at 45 degrees from the sun. The technical name for this phenomena is Parhelia. The corresponding image in connection with the moon is called Paraselene. Another startling apparition in conjunction with the sun is called a sub-sun. Sub-suns result from the reflection of the sun in a layer of flat ice crystals and appears at a point below the real sun and can be as brilliant as the sun itself. The sub-sun can develop a pattern of sun dogs and haloes causing a further complicated illusion. At night, the moon will reflect in the same manner under like meterological conditions. This type of apparition is particularly discernible from aircraft at high altitudes.

Cirrus cloud formations are effective viewing screens for illusions resulting from reflected or refracted light, as they contain ice crystals. These clouds exist in the upper atmosphere, so that conditions are favorable throughout the year for sun dog and moon dog apparitions. However, such phenomena usually are discernible at lower levels only during winter months in temperate zones.

The aurora borealis, or northern lights, produces conditions and phenomena which have been associated with mistakenly conceived flying objects. Auroral activity is associated with the earth's magnetic fields, explosions on the surface of the sun, and other solar activity. The auroral zone in the northern hemisphere follows roughly a circle around, and about 23 degrees away from the magnetic pole. In Europe, auroras are seen only infrequently below 50 degrees.

The aurora borealis cannot be seen in full daylight, and during moonlit periods it is inconspicuous. It is sometimes bright enough to read by, and on rare occasions, its surface brightness surpasses even that of the moon. The most distinctive form of the aurora is that of a curtain or long wavy band, often with folds and flutings in it. Although the lower edge of the aurora is nearly horizontal, the band as seen from Europe would appear as an arc, due to its great distance from the observer. Auroras may consist of more than one curtain and may appear and disappear rapidly, remain constant for long periods, or move slowly across the sky. Some may appear merely as formless, diffused lighting in the sky. Faint auroras may appear colorless. Bright auroras are usually yellow-green, but other colors such as red, blue, grey, and violet sometimes appear. A yellow-green curtain often will be tinged with red around its lower edge. Auroras may appear high in the sky or low on the horizon, depending on the distance of the particular phenomenon from the observer.

While the chances of the aurora borealis itself being mistaken for a flying object are remote, the erratic lighting conditions it produces may often be a contributing factor to a sighting.

There are other phenomena believed to be associated with auroral activity which can produce apparitions resembling flying objects. Such phenomena occur during magnetic storms and probably are the result of gases emitted from explosions on the sun, and other solar activity. One such phenomenon, observed in northwest Europe, was described as a large brilliant disk which appeared on the east-northeast horizon and moved slowly across the sky, changing into an elongated ellipse, thence back to a disk before it disappeared below the opposite horizon.

This phenomenon was observed by many scientists who were out in force to observe expected auroral displays in connection with the magnetic

It's Easy to Be Fooled

storm they knew to be in progress. It is believed to have been caused by gases travelling through layers of the upper atmosphere in the auroral zone. Its color was described variously as white, pearly-white, greenish-white, and yellowish-white. Calculations based on numerous observations of the phenomenon indicate that it may have been about 70 miles long by 10 miles in diameter.

This phenomenon occurred before the advent of the airplane and all observations were from the ground. However, a phenomenon of this size and brilliance could be seen for hundreds of miles from the air, and in myriad fantastic shapes and maneuvers if complemented by compatible atmospheric conditions. Official astronomical records reveal numerous equally fantastic illusions resulting from phenomena of this sort.

The composition and structure of the earth's atmosphere and the space which lies beyond, and the natural laws which govern them, are complex.

The foregoing is not an attempt to relate all apparently unexplainable aerial phenomena to meteorological and astronomical causes. Rather, it is a summation of the more important aspects of meteorology and astronomy which contribute to sightings of illusionary and real flying objects that cannot be identified readily. The information is designed to orient the potential observer in meteorological and astronomical conditions which affect human perception, thereby enabling him to understand the implications involved and report his sightings more rationally and lucidly.

In certain instances, unidentified objects have been observed on radarscopes, both ground and airborne. Generally speaking these radar sightings fall into explainable patterns and are caused by certain meteorological phenomena, or familiar objects that are observed under unnatural circumstances.

Radar echoes can be produced by a variety of objects, many of which are not visible to the human eye. A majority of solid objects which return radar energy produce responses on the radarscope that are easily recognizable. Moving objects, such as birds, aircraft, and meteorological balloons, are normally recognizable by their size and velocity. However, some balloons, such as ionospheric balloons, ascent to altitudes above those of normal aircraft and travel with the upper air currents, sometimes at speeds above 150 miles per hour. Radar returns from these balloons could give impressions of unidentified objects.

Certain meteorological and astronomical conditions will present radar

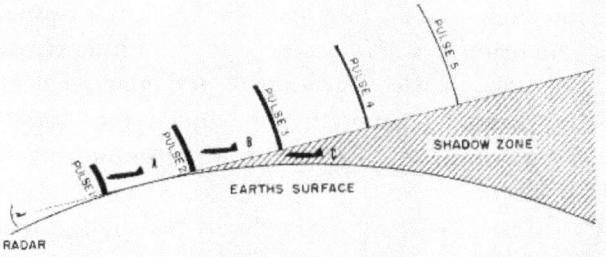

The transmission of a radar pulse, under normal atmospheric conditions, follows line of sight. Therefore the curvature of the earth would place Target "C" in the shadow zone.

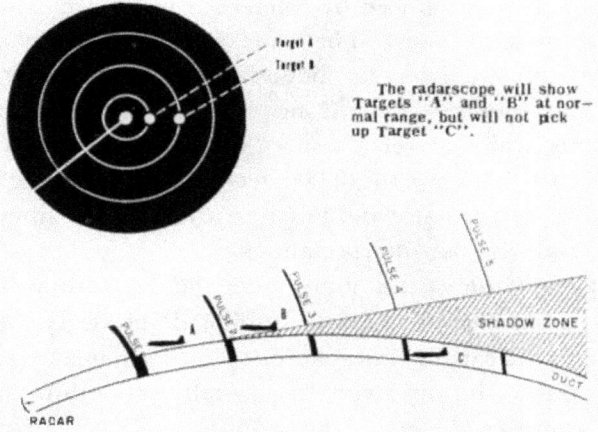

The radarscope will show Targets "A" and "B" at normal range, but will not pick up Target "C".

Under abnormal conditions, with cool air overlaid by a warmer air mass, a duct is formed through which the radar pulse travels and reflects Target "C" at a much greater distance.

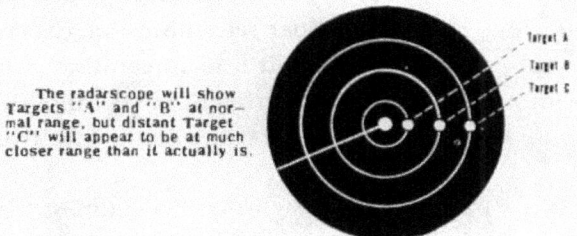

The radarscope will show Targets "A" and "B" at normal range, but distant Target "C" will appear to be at much closer range than it actually is.

A temperature inversion produces radar phenomena.

It's Easy to Be Fooled

returns that are unusual. Radar waves must travel through the earth's atmosphere where, like light waves, they may be bent by unusual temperature and moisture conditions. Radar waves may be refracted or reflected by atmospheric conditions to where ground objects may seem to represent an aircraft or flying object. Even with a moving target indicator, reflected images of distant ground objects may appear to be moving because of the movement of air layers.

Temperature inversions, in which a cold air mass is overlaid by a warmer air mass, can greatly increase the distance from which normal radar returns are received. Thus, objects may appear to be much closer than they actually are and these distant objects, superimposed on the normal radarscope picture, may result in misinterpretation and confusion.

Radar echoes may be produced by condensed water vapor in the form of raindrops, ice crystals, or snow. These radar reflections may cover a wide area which has diffused, irregular boundaries and fluctuating intensities. Movement of this water vapor will be determined by the movement of upper air currents, which travel at a speed of as much as 100 miles per hour or more and at altitudes up to 40,000 feet. Normally, these patterns are easily recognizable by their size and radar return; however, they may appear confusing and result in false interpretations.

Meteors that enter the earth's atmosphere and get within range of radar may cause reflections that are extremely difficult to verify. Meteors reach the outer fringe of the earth's atmosphere at a rate of something near 100,000 miles per hour, although only a very few actually get within range of radar. Those that do, approach the earth from all angles and at velocities approaching 25,000 miles per hour. Radar responses to these meteors may occur at any range or altitude, depending only upon the capabilities of the radar set. Radar reports resulting from this type of phenomenon can be verified by a study of the expected paths of meteors at the time of the incident.

In addition, there is the possibility that one radar set, which has characteristics similar to those of another radar set within range, may cause interference and unusual responses that could lead to confusion and inaccurate interpretation. Although this type of interference may cause the appearance of one or even two targets on the radar screen, it can generally be recognized quite easily.

A careful study of unusual radar sightings will almost always disclose that the reason is explainable. Experience in the operation of radar will provide the operator with the ability to recognize most unusual phenomena when they occur. However, occasionally a vertification of meteorological or astronomical data may be necessary to substantiate the validity of what otherwise might be considered an unfamiliar flying object.

Physiological factors may have profound effects upon an individual's ability to observe and to interpret observations accurately. One of the greatest hindrances to human understanding can result from deception of the senses. The sense of sight is, by itself, purely a physical process and the perception and understanding attached to visual sightings is determined largely by memory of past experiences and familiarity with surrounding objects. This relation of experiences to the interpretation of visual sightings permits many errors.

This margin of error may be applicable particularly to aircrew members operating high-performance aircraft, under adverse or unusual weather conditions, under tension, and during periods of extreme fatigue.

The aircrew member is generally familiar with many of the unusual observations associated with meteorological and astronomical phenomena. However, many unusual observations are the result of certain physiological effects that may be unknown or unfamiliar.

Occasionally, objects that exist on the surface of the eye may be mistaken for distant objects. These objects take various forms. Tiny specks of dirt may appear as shimmering globules of light and if a speck is illuminated by an outside light source, it may appear as a large, out-of-focus blob of light. If this speck is viewed against a dark sky or background, it may be quite spectacular. As this speck floats across the pupil of the eye, it will create the appearance of movement.

Many reported unidentified objects, described as flying saucers, flying disks, shiny spots, or a string of pearls, are nothing more than minute blood capillaries on the surface of the retina of the eye, or tiny corpuscles, which become visible under special conditions of illumination.

Another physiological phenomenon is that of after-image. A sudden flash of light affects the retina of the eye and causes a dark image to remain visible for some time after the light has been extinguished. Flashes of lightning, comets, or meteors will cause this effect and may be confused and interpreted as unidentified flying objects.

Hypoxia, resulting from lack of oxygen, has varying effects on the ability to react and to observe accurately. The effects of hypoxia may vary much in the same manner as those of alcoholic intoxication. Usually vision is affected, reactions are retarded, and observations are distorted. An oxygen

It's Easy to Be Fooled

mask leak may cause alternating stages of hypoxia and normalcy, with the individual often being unaware of these changes.

In a series of tests conducted at the Air Force School of Aerospace Medicine to determine the effects of fatigue, it was discovered that extreme fatigue may cause an individual to hallucinate, imagining that he sees a variety of unusual objects, and with a vividness to make them seem quite real. Fatigue, even in minor degrees, will slow down reaction time and reduce ability to observe and interpret observations.

Two phenomena that occur frequently are those of autohypnosis and autokinesis. In both of these reatcions, a stationary light will assume apparent movement. In autohypnosis, this reaction is caused by continued attention to an external light source. Autokinesis is the result of observing a stationary light under circumstances in which relation to familiar objects is absent.

There is strong evidence that a great many visual problems, both physical and physiological, arise as a direct result of flight at high altitudes.

When flights are conducted at relatively low altitudes, the visibility of distant targets will be reduced by atmospheric haze. This is because light emanating from objects in space is gradually reduced by absorption and by primary and secondary scattering along the pathway of sight.

Along with the variation of the contrast by atmospheric interference, there is a shift of the apparent contours. This has been disclosed by experiments performed at the Air Force School of Aerospace Medicine. From these studies, it was concluded that the apparent angular size and apparent distance of objects depend on the brightness reduction of the atmosphere. With increasing altitudes, the deviation of the apparent luminance from the actual luminance of an object in space will result in the object's appearing brighter than it actually is. This may result in false identification of a normally familiar object.

The physiological effects enumerated above are but a few of the manifestations resulting from known reactions. Many physiological effects resulting from high-performance flights are still in the category of unknowns. However, these factors greatly influence one's ability to understand and interpret sensory actions. If recognized by the aircrew member, they may aid in identifying unfamiliar objects in flight.

Reasoning ability, degree of susceptibility to suggestion, and general mental attitude are vital factors in identifying and reporting flying objects. Failure to note details accurately and a tendency to overdraw descriptions of sightings can result in failure to identify. An over-active imagination, coupled with physiological strain, can transform unfamiliar meterological or astronomical phenomena and light aberrations into unidentified flying objects.

Perception and feeling are closely related and can have a marked effect upon understanding. Motivation in many instances determines how we interpret what we see, and expectancy can induce manifestations which are only indirectly related to actual physical phenomena or objects. The separation of what may be observed through the senses from what is known through thought or intuition is difficult, inasmuch as understanding is derived from a combination of both. However, an objective attitude, which permits assessment of observed characteristics, rather than suppositions or theories, will assist the observer in avoiding distorted descriptions.

It has been suggested that the world each of us knows is a world created by large measure from our experience in dealing with our environment. When two points of light, one brighter than the other, are placed at an equal distance from an observer in a dark room, the bright point of light looks nearer than the dim light, if one eye is closed and the observer remains motionless. The direction from the observer, as well as a difference in brightness, will result in an apparent variance in distance. Should two equally bright lights be placed near the floor, one about a foot above the other, the upper light will appear to be at greater distance from the observer than the lower one. Conversely, when the lights are placed near the ceiling of the room, the lower light will appear to be farther away.

When two partly inflated balloons are illuminated indirectly and fastened in positions about one foot apart, where their relative brightness and inflation can be controlled, the observer will experience a variety of reactions as to what he saw.

If the brightness and size of the two balloons remain the same and the observer views them with one eye at a distance of approximately ten feet, he sees two bright spheres equidistant from his position. If the relative sizes are changed and the brightness remains the same, the larger balloon usually appears nearer. When the size is changed continuously, the lighted balloons seem to move back and forth, giving the effect of erratic movement of lighted spheres through space. This is true even when observed with both eyes. If the relative brightness is varied constantly and the size remains the same, a similar effect is obtained. When there is a variation in relative size and brightness, most observers are inclined to judge distance by relative size rather than by relative brightness.

It's Easy to Be Fooled

The effect of these tests upon the observer is premised on the fact that he draws upon past experience in assessing distance based upon relative size and brightness. He assumes that, since the two points of light appear similar, they are identical and of equal brightness. Therefore, the point of light which seems brighter must be nearer. In the case of the two points of light placed one above the other, past experience leads the observer to assume that, when he looks down, the lower light is nearer and, conversely, that, when he looks up, the higher light is nearer.

With regard to the seeming variance in distance when the size of objects is changed continuously, rarely has the observer seen two fixed objects at the same distance change in size. Usually any change in size of an object results from a change in the position of the object in relation to the position of the observer. As the object draws nearer, it becomes larger, and the reverse is true as it draws farther away. Therefore, in the case of the two balloons, the observer assumes that any change in size of the two balloons results from a variation in distance from his point of observation.

These experiments show how misinterpretations can result from the relation of visual perception to past experience in an effort to understand and recognize the object or objects seen.

When we see an object, we derive an impression not only of its location, but also of its existence as an object, and the location as related to visual perception will color the characteristics it possesses. Objects seen through haze or mist, or in reflected light, will assume characteristics they do not possess normally, but, because they have been perceived visually, the observer tends to accept them as real. Thus, psychologically, he creates an object with characteristics which do not exist in actuality. It is essential, therefore, that the observer analyze his observations in relation to unusual weather or lighting conditions and reject characteristics which deviate from the normal and can be explained by the unnatural conditions under which they were seen.

When we see an unfamiliar object, we draw upon our individual past experience in an attempt to identify it. If the unfamiliar characteristics of the object cannot be related to past experience, we have a feeling of uncertainty and it is then that we draw upon imagination in an effort to relate visual perception to understanding. Imagination is colored by suggestion and herein lies an inherent danger.

We are open to suggestion constantly in our daily lives. Advertising media, artists' concepts, modern-day science fiction, propaganda, exaggerated films, publicity on perpetrated hoaxes, and the imaginings of zealots and fanatics all react upon the consciousness in the form of suggestion. When we seek an explanation for the unusual or unfamiliar, and attempt to draw upon imagination instead of rationalization, suggestion influences our thinking.

Physiological changes due to fatigue and intense strain enhance the susceptibility to suggestion and may induce psychological manifestations which a more rational state of mind would reject. The observer should attempt to evaluate his observations. Objective analysis of those characteristics he has observed, in relation to the conditions under which they were seen, will assist in identification of the unfamiliar object and result in more accurate reporting.

It is a common misconception that the eye "takes a picture" of everything within its field of view. This is not true. Pick out any word in this sentence and then move your eye to the next and then the next. You will discover that you can no longer read the first word after having moved your eye about 5 degrees.

You see best in daylight and the eye sees by moving in short jumps. It is not a sweeping but a jerking motion with which you see details around you. This is of the utmost importance to the combat pilot scanning the sky for the enemy. Experiments have shown that the eye sees nothing in detail while it is moving. It sees only when it pauses and fixes an object on its retina. In scanning the sky, do not deceive yourself that you have covered an area with a wide, sweeping glance. The correct way to scan is to cover an area with short, regularly-spaced movements of the eye.

Judgment of distance is done subconsciously in a combination of ways: Close up, we depend on binocular vision, each eye seeing an object from a different angle. At distance beyond binocular range, which is usually the case in flight, we judge it on a one-eye basis.

The eyes change focus to see objects within about 20 feet, but the change in focus for distant objects is negligible.

It is easy for your eyes to play tricks on you at night when you stare for some time at a light—say, the tail-light of an automobile or a lead airplane. What happens is technically known as autokinetic movement, or more commonly as stare vision. If the light is stationary, it may seem to move and swing in wide arcs. If the light is moving, it may seem to move to the side when it is actually going straight ahead. The cure for stare vision is don't stare—keep shifting your gaze from point to point.

Another common illusion at night is to see a light expanding or contracting at a fixed distance from you when actually the light is approaching or going away. Again, shift your gaze.

It's Easy to Be Fooled

Dust, grease, water droplets, scratches, on the windshield all obstruct vision, night or day. A speck on the windshield could, after a few hours, take on the silhouette of "an unidentified flying object."

With regard to color perception at night, blue and green lights are seen most easily; red and orange are seen least easily.

There is much room for an error in judgment in seeing unidentified flying objects. This is normal and can happen to anyone.

6
LISTEN TO THE EXPERTS

In reporting thousands of UFO reports there have been many observers referred to as experts because of some technical training in a highly specialized field relative to aviation and air travel. Military pilots and airline pilots, ground radar operators and airborne radar operators, aerial navigators, engineers and ground observer corp members are among these. To label these people as experts in this field of UFOs and sky phenomena is absolutely wrong. These people are, of course, highly qualified to do their specified jobs and the fact that they may have observed an unidentified flying object does not in any sense of the word make them incompetent or inadequate in their particular field of endeavor.

People from all walks of life have seen and have reported UFOs to the United States Air Force. The great majority of them are not really experts in astrophysics, or visual astronomy. Visual astronomers, astrophysicists, and their associates are the only ones who truly qualify as experts in this field of UFOs.

Let's see what some of these people have to say on this subject.

During a TV presentation entitled "UFO, The Enigma of the Skies" presented by the Armstrong Circle Theater on January 22, 1958, Professor Donald H. Menzel, Professor of Astrophysics and Director of the Observatory at Harvard University was interviewed by Mr. Douglas Edwards as follows:

> Professor Menzel . . . in your considered opinion . . . has our atmosphere been invaded by spaceships . . . flying objects . . . originating in outer space?
> (MENZEL): No. An emphatic no! Now I have no quarrel with the assumption that there is life on other planets. As I have stated on a number of occasions, "life . . . even human or superhuman life, may exist in millions of places in the universe." And I also concede that space travel for earthlings is imminent. But these two facts cannot justify the conclusions that flying saucers represent space travel in reverse.

Listen to the Experts

(EDWARDS): And yet . . . tonight . . . we've heard a good bit of evidence that UFO . . . unidentified flying objects . . . do exist . . .

(MENZEL): They are "unidentified" flying objects only because of mistaken identity. In fact, they are often not even objects. That is why I prefer the name "Flying Saucers," rather than UFO, which implies that they are material or tangible. The observations, and most particularly the . . . interpretation of these observations . . . is strictly a scientific matter. You wouldn't call in a carpenter to diagnose a pain in your stomach . . . why call in untrained laymen when you want to diagnose a strange occurrence in the sky? And strange things do occur . . . all the time. I think Project Bluebook hasn't paid enough attention to aerial phenomenon in its analysis work. Mirages, ice-crystal reflections . . .

It's understandable that untrained observers are startled when they see unusual occurrences in the sky . . . it even happens to many of us who have studied the heavens for years. It happened to me. Let me give you just the bare details. It was three years ago, I was flying from the North Pole to Point Barrow. Suddenly I saw an object pop over the horizon and buzz my plane twice. Then it stopped and flew along parallel to it at a distance of about 300 feet. I calculated it to be about two feet in diameter, it seemed to have flashing red and green lights and something that looked like a lighted propeller on top.

(EDWARDS): Did you ever find out what it was?

(MENZEL): Certainly . . . as soon as I got over the initial shock I did some calm analyzing. The mysterious saucer proved to be a mirage of the bright star, Sirius. Actually, the star was slightly below the horizon . . . but the queer bending of light by the atmosphere known as refraction had brought it into view. Distortion and diffusion caused by alternate layers of warm and cold air had added the finishing touches, creating a bizarre, pulsating, whirling effect similar to the twinking of a star. That was the answer to my saucer . . . and to a lot of other people's saucers, I'd like to bet. And I've seen hundreds of other saucers from dozens of causes. For instance, high flying spiderwebs, and owls that glow in the dark.

(EDWARDS): In other words, if more people were astronomers, there would be fewer Flying Saucers.

(MENZEL): Fewer Flying Saucers and a lot less beclouding the issue by over-eager amateurs. . . .

(EDWARDS): Well, let's agree that astronomers are best equipped to evaluate aerial phenomena . . . but wouldn't you agree that pilots, meteorologists are also trained observers . . . and many of them have reported UFOs.

(MENZEL): No, they're not trained observers in this sense. A pilot can observe a plane and know what it is. A meteorologist can spot a cloud and not be fooled . . . but many, many things occur in the sky that are completely beyond the knowledge of experienced pilots or weathermen. Mirage . . . mock suns caused by reflections from ice-crystals . . . are common. But when they do happen, only trained scientists can recognize and understand them. If a pilot swerves his plane to avoid a fireball that misses him by hundreds of miles, or if he fails to identify a sundog, that doesn't mean he isn't competent as a flyer . . .

(EDWARDS): Well, summing up then, you feel that all the Flying Saucers reported in the past 10½ years . . . even before that and even including the 1.9% the Air Force classified as unknown . . . can be explained as completely natural occurrences . . .

(MENZEL): I most cetainly do. And I think that the Air Force has been put upon in this whole affair. They've had to make their investigation of course . . . they're bound by the most serious of considerations: the security of our country. But it's gone far enough. In my opinion, it's not the reports of Flying Saucers that should be analyzed . . . it's the non-qualified interpreters themselves who argue that these saucers come from outer space.

On the same program, the Honorable Richard E. Horner, then Assistant Secretary of the Air Force for Research and Development said:

During recent years there has been a mistaken belief that the Air Force has been hiding from the public information concerning Unidentified Flying Objects. Nothing could be further from the truth. And I do not qualify this statement in any way. The activities of the International Geophysical Year, which is serving to focus more trained eyes on the skies than ever before, climaxes 10 years of Air Force investigation of UFO sightings. All but a small percentage of these reports have been definitely attributed to natural phenomena that are neither mysterious nor dire. A report of a special scientific panel on UFO, assembled at the request of the United States Government, concludes that the evidence shows no indication that these phenomena constitute a direct physical threat to national security. They find no evidence that UFOs are capable of hostile acts. They find no evidence that UFOs indicate a need for revision of current scientific concepts. And finally, they recommend that the subject be stripped of the aura of mystery it has unfortunately acquired. The panel's findings coincide completely with the announced position of the Air Force. There is no evidence at hand that objects popularly known as "Flying Saucers" actually exist. Not a single speck of material evidence, not a fragment of a "Flying Saucer" has ever been found. This does not mean that the Air Force can or will be complacent about UFO reports. We will continue our program of detailed investigation of all UFO sightings—and invite everyone who observes an object in the sky he sincerely cannot identify or explain, to report it immediately to the nearest Air Force activity.

Listen to the Experts

Mr. Horner formerly Assistant Secretary of the Air Force for Research and Development and Associate Administrator for the National Aeronautics and Space Agency is now Senior Vice President (Technical) of Northrup Corporation.

Another interesting and recent development is that Captain Ed Ruppelt, author of The Report On Unidentified Flying Objects, and an analyst for the Air Force's Project Bluebook, who associated with many astronomers and astrophysicists, recently revised his book by adding three chapters. He sums up his belief by saying "even taking into consideration the highly qualified backgrounds of some of the people who made sightings there was not one single case which upon the closest analyses could not be logically explained in terms of some common objects or phenomena."

Dr. Addison M. Duval, Psychiatrist and Deputy Director of Washington's famous St. Elizabeths Hospital, recently told the United Press International that seeing things which don't exist is a common result of anxiety generated by fear of the unknown and added times aren't getting any less anxious or the future any less uncertain. Dr. Duval stated that he thought flying saucers were going out of fashion. If the flying saucer delusion is on the decline, one reason may be that saucers are getting "too much competition from real things—satellites, lunar probes, sun rockets," Duval said.

Dr. Hugh L. Dryden, Director of the National Aeronautics and Space Agency states emphatically that there is no such thing as a flying saucer.

"It just isn't so," declared former General James H. Doolittle, formerly head of the NACA, "there are no flying saucers."

Dr. Peter van de Kamp of Sproul Observatory, Swarthmore College, on August 25, 1959 stated "I am an astronomer, have never observed a bona fide 'flying saucer,' and have the impression that there is a strong correlation between the incident of 'flying saucers' and the state of mind of an observer, influenced by fear, hysteria, wishful thinking, or plain fraudulent intentions."

Dr. Otto Struve of the University of California in a letter dated July 21, 1959 stated "I have never observed any phenomenon in the sky or read of the descriptions of phenomena observed by others that would inspire confidence in a hypothesis of an extra-terrestrial origin of the so-called flying saucers." This from a world renowned scientist, director of the National Radio Astronomy Observatory who announced on April 26, 1960 that his group would investigate the possibility of life on other planets by listening for intelligent signals from space.

Dr. William A. Calder, Professor of Astronomy, Bradley Observatory, Agnes Scott College, Decatur, Georgia, in a letter dated July 21, 1959 states "While many strange sights have been reported in this area, all flying saucers and visitors from outer space have given Bradley Observatory the complete brush-off. What is it about us that they don't like?"

Willy Ley, pioneer in rocket research and space travel says in his book *Satellites, Rockets and Outer Space,* "the flying saucer epidemic seems to have run its course in the same manner as many epidemics caused by bacteria. At the first onslaught a large portion of the population succumbs, getting more or less severe cases of the disease. A few are immune and are treated with sneers by those who have caught it; immunity seems to be unfair in an undefinable manner. After a while most of the patients recover but there are a number of hopeless cases."

Dr. J. Allen Hynek of Northwestern University and the Smithsonian Institute says:

> The universe, as revealed by the researches of astronomers, is so vast and comprises so many billions of suns—and hence by inference, billions of solar systems too—that for one to hold today that man is the highest intelligence in the universe exposes him to the charge of cosmic provincialism and egocentricity.
>
> Granted, then, that there may be countless other intelligent civilizations. Granted also that they might attempt to communicate with other such civilizations, ourselves included. But here is where the cosmic romantic and the serious astronomer must part company. The very vastness of the universe places a virtually insuperable obstacle to the fulfillment of expeditions of communication. The distances involved are simply too great and the engineering difficulties too fabulous for us to conceive of such visitations.
>
> But our wishful thinkers come back and say, "But what about all the evidence of flying saucers that the Air Force has been gathering all these years. Let the astronomers alone, let's just look at the evidence." Ah, there's the rub! What evidence? As consultant on these matters to the Air Force for many years, I have seen this evidence; and as a scientist it leaves me quite frustrated. There is nothing there that any scientist would truly call scientific evidence. There are no detailed numerial results that can be used as the basis of computations. There are no spectrum analyses of light, no photographs that reveal any detail—generally nothing but vague statements that lack the all-necessary scientific precision. Almost all of the reports furthermore demonstrably arise from stimuli which can most logically be identified with misidentification of some object—balloons, birds, aircraft, etc.; and what is most frustrating and disappointing to me,—shall I confess it—I should very much like to be one of those cosmic romantics who believes in

visitors from space. How exciting! What new meanings in philosophy and religion! But, alas, I must remember first that it is not personal belief but hard, cold facts presented by the evidence that are the building bricks of science. So far—and much to my disappointment—the evidence has been found wanting. Even in those few cases the Air Force cannot explain, and of which I have investigated a number personally, and I cannot explain, either the scientific raw material is insufficient to come to any definite conclusion or even those famous "unknowns" do not appeal to my reason as the sort of things by which distant intelligent beings would try to communicate with us. Why be so mysterious? Why appear in such trivial ways and to so few? Why not present a full-scale display to a full city?

To astronomers the claims of the flying saucer proponents do not make sense scientifically—as yet anyway. But I don't think there is an astronomer alive who would not stand up and cheer at some real evidence for space visitors. How our knowledge of astronomy would grow if we could communicate directly with denizens of outer space and how our appropriations would mount! And Congress would certainly grant astronomers and the Air Force vast funds to pursue such communications! From that standpoint alone I might well say, "Please let us have some real flying saucers!"

And in a more recent letter, dated March 29, 1960, Dr. Menzel says

As you know, in 1953 I published a book, "Flying Saucers," by the Harvard University Press. This book was a scientific study of the limited observational data then available including a number of special cases released by the Air Force. Although my conclusions were then based largely on theoretical considerations, I have during the past seven years had an ample opportunity to check and re-check the statements that I made in that book. Although I would not claim 100 per cent accuracy in solving all of the cases, the material that I had available at that time was consistent with the solutions that I then proposed. I have seen mirages that have actually 'buzzed' an airplane. I have seen spectacular wingless vehicles suddenly materialize as if by magic and pace a ship in its path through space. I have seen some of the famous green fireballs of the Southwest, including one that has been widely quoted in the flying saucer literature.

I have, in my opinion, duplicated or seen duplicated practically every famous flying saucer apparition in the books. Every one has proved to have a simple solution, in terms of reflections or illuminations from natural objects. Or mirages, refractions from ice crystals, reflections from water droplets, and so on have accounted for the rest.

I adhere strongly to my original proposition. That flying saucers have a completely natural explanation when sufficient data are available. There was absolutely no evidence whatever to support the contention of certain writers that flying saucers come from outer space. Furthermore, I have had sufficient contact with the Air Force, during recent months, to confirm my previous conclusions, namely that there is no evidence in the Air Force files tending to support the sensational conclusion that flying saucers are from outer space. Nor are they in any sense a menace to the United States as they would be if some foreign power had engineered them.

And Dr. Jesse L. Grenstein of Mount Wilson and Palomar Observatory, Carnegie Institution of Washington, California Institute of Technology, on July 23, 1959 made the following perceptive argument:

> If we assume that many brilliant objects of unknown origin move rapidly in the earth's atmosphere, and are easily visible to the naked eye, from planes or from ground level, it is remarkable that none have, by accident, been seen or photographed by professional astronomers. Inspection of the entire sky many times a night by hundreds of astronomers all over the world has been going on for several hundred years. Comets are searched for on regular programs by both amateur and professional observers; about ten faint comets, most too faint to be seen by the naked eye, are found each year. No reputable astronomer has reported large, bright, and rapidly moving objects except for fireballs and meteors, in spite of this systematic coverage of the sky. Observatories are generally placed in areas of clear sky. It might be objected that flying saucers would attempt to avoid detection, but some observatories are located near big cities where sightings have been reported and thousands of amateur astronomers and observers exist scattered through the country.
> The entire night sky is photographed on a systematic and regular basis many times a year. For example, a collection of half a million photographs taken with wide-angle, fast lenses exists at the Harvard College Observatory. These telescopes photograph, in about an hour's exposure, stars a hunderd times fainter than can be seen with the unaided eye. These record numerous meteors, light tracks produced by tiny dust particles plunging into the earth's atmosphere, at ranges of 50 to 100 miles. Yet none of these photographs show a bright, rapidly moving, large object such as "saucers" are reported to be. The wing-tip lights of planes are frequently recorded on such astronomical photographs, in spite of the faintness of such a light source compared to the reported "flying-lights."
> A complete survey of the sky was completed over a period of five years at the Palomar Observatory. The sky was photographed in part on over 1000 nights. About 5000 plates were taken with a telescope of 48-inch aperature, which has a field-of-view of 50 square degrees. Stars 100,000 times fainter than the naked-eye limit were photographed by millions. Hundreds of meteor

Listen to the Experts

trails and wing-tip lights were photographed. No mysterious, large, bright and rapidly moving object is recorded. When such a telescope is pointed at a low angle above the horizon, it looks through a column of the earth's atmosphere 50 to 100 miles long and might be expected to pick up any object within that range. A luminous moving object 100 feet in diameter, at a distance of fifty miles would leave a track on the plate over a millimeter wide and could be easily distinguishable from a meteor, or star, which has images 30 times smaller. No such objects have been seen.

Admittedly, this is a small cross-section of experts with the large group of astronomers who must scan the entire sky many times a night, it is significant that none of these aerospace experts have witnessed any evidence of interplanetary spaceships.

It is also significant that a number of these experts feel that the flying saucer era is coming to an end. Technological advances by the United States and Russia heralding imminent space travel for man, have already unlocked some of these secrets of outer space. Two of the most notable being the Russian photograph of the backside of the moon and the United States orbiting Pioneer V satellite around the sun.

Therefore, admitting the possibility of the existence of some form of life within our solar system, or in our own galaxy, as any true scientist would, and based upon our recent advances in space technology, it seems evident that earthmen will visit these faraway neighbors before they can visit us.

7

THE OFFICIAL AIR FORCE POSITION

As General Thomas D. White has indicated previously in this book, the Air Force investigation and analysis of UFOs over the United States are directly related to its responsibility for the defense of the United States. Because prompt reporting and rapid identification are necessary to carry out the second of the four phases of air defense, detection, identification, interception, and destruction, the Air Force maintains the unidentified flying object program.

Statistics for the past 13 years are as follows:

YEAR	NUMBER OF OBJECTS SIGHTED AND REPORTED TO THE USAF
1947	79
1948	143
1949	186
1950	169
1951	121
1952	1501
1953	425
1954	429
1955	404
1956	778
1957	1178
1958	573
1959	364
1960	173
TOTAL	6523

In analyzing the above figures the reader will quickly see that the Air Force's peak years for UFO reported sightings were 1952 and 1957. Further analysis shows that there are usually definite reasons or specific incidents causing a peak year or a rash of sightings. In the year 1952 the famous Wash-

The Official Air Force Position

ington sightings discussed in Chapter III touched off a rash of sightings for the last six months of that year. In 1957, Sputnik I, the Russian satellite was launched in October. Seven hundred and one of the total sightings for that year took place in the last three months of the year. This is most significant.

In all the correspondence received by the Air Force over the past thirteen years there are a number of allegations and charges against the Air Force which are repeated time and time again and which should be specifically answered.

> The first charge states that there is a document dated September 23, 1947 which is a Secret conclusion by the Aerospace Technical Intelligence Center that the flying saucers were real.
>
> The second charge states there was a 1948 Top Secret document concluding that the unidentified objects were interplanetary spaceships.
>
> The third charge states that a Secret Air Force intelligence analysis of UFO maneuvers concludes that the objects are interplanetary.
>
> The fourth charge states that a Secret report by a panel of top scientists convened in the Pentagon in January 1953 urges that the Air Force quadruple its UFO project and that the Air Force give the American people all UFO information, including Secret Air Force conclusions, unsolved sightings and photo analyses.

In answer to the first charge, there is no official Air Force report or document which states that the so-called flying saucers are real. The writer does not doubt that in the early days of the UFO program many possibilities, including this one, were probably listed in order to develop an adequate investigation and evaluation program for the Air Force. Also, it is conceivable that some person or persons associated with the Air Force program were personally convinced that flying saucers might be real and could be interplanetary spaceships.

In answer to the second charge, there has never been an official Aerospace Technical Intelligence Center estimate of the situation which stated that so-called flying saucers were interplanetary spaceships. Again, individuals associated with the early Air Force program, listing all possibilites, would probably have consdered such a category but such a conclusion or document reaching such a conclusion is non-existent.

The answer to the third charge is that such a report or intelligence analyses is non-existent. There is no such report. Again, there has never been an Aerospace Technical Intelligence Center conclusion that any of the UFOs were interplanetary.

The answer to the fourth charge is a simple one. On January 14, 15 and 16, 1953, at the request of the United States Air Force, a scientific advisory panel was established to consider the UFO program and to make recommendations based upon their conclusions. The recommendations of the panel as reported in the fourth charge above are completely erroneous. The final conclusion reached by the panel since declassified and released to the public were that UFOs (a) held no direct physical threat; (b) were not foreign developments capable of hostile acts against the U.S., (c) were not unknown phenomena requiring the revision of current scientific concepts, and (d) the panel further concluded that unless de-emphasized, UFOs, or the subject itself, could constitute a threat to the national security because a rash of sightings could effect defense communications, national hysteria could be induced by skillful hostile propagandists, and a mass of false reports could screen planned hostile actions against the United States. As a result of this 1953 meeting, the panel made the following basic recommendations:

(a) That immediate steps be taken to strip the UFOs of the aura of mystery which they had unfortunately acquired;
(b) that the public be reassured of the total lack of evidence of inimical forces behind the phenomena,
(c) that Air Force investigative personnel be trained to recognize and reject false indications quickly and effectively.

An additional charge against the Air Force is that it is afraid to tell the public the truth concerning flying saucers because of the national hysteria and panic which will result. This is, of course, a ridiculous argument because the same groups hurling these charges and allegations against the Air Force are continually telling the public the same thing they want the Air Force to say and there is no resulting panic. The real truth of the matter is that the public itself trusts our government and our Air Force and does not believe that space people are visiting our planet in extra-terrestrial vehicles.

Reporting, investigating, analyzing, and evaluating procedures have improved considerably since Kennth Arnold's sightings of the flying saucer on June 27, 1947. The study and analysis of reported sightings of UFOs is continued today by selected scientific personnel under the supervision of

THE OFFICIAL AIR FORCE POSITION

the Air Force and by qualified scientists and engineers from outside the Air Force on a contract basis.

Dr. J. Allen Hynek, Professor of Astrophysics at Northwestern University and Associate Director of the Astrophysical Observatory for the Smithsonian Institute, Cambridge, Massachusetts is the Chief Scientific consultant to the Air Force on the subject of UFOs. In this capacity Dr. Hynek presided at a committee meeting held by the Department of the Air Force in the Pentagon on February 17, 1959 for the purpose of discussing the Air Force philosophy toward and policy regarding the unidentified flying object program.

Recommendations resulting from this meeting were as follows. The Air Force must continue to take a positive approach toward the UFO program. Air Force investigations of reported sightings must be thorough and scientific with all possibilities considered. The entire resources of the Aerospace Technical Intelligence Center must be made available to help with the overall analysis and evaluation program of reported UFOs and the *public must be kept fully informed* of the Air Force position regarding UFOs. These recommendations have been carried out and today a special Air Force UFO committee meets each month to make sure that every means available is being used in pursuit of a positive UFO investigation program and that a thorough information program is being conducted to keep the public informed.

The official Air Force position can be summed up in a few words. In its attempt to put the UFO subject in proper perspective, it appears that the Air Force cannot compete with the science-fiction writer in satisfying the desires of those people who wish to believe in spaceships. The Air Force believes that the investigation of the UFO phenomena is in responsible hands and that an adequate, thorough, and honest program is being conducted. Regardless, a small but articulate segment of people are under the mistaken belief that the Air Force has not sought the assistance of outstanding authorities from outside the Air Force to assist in its evaluation of UFO sightings and that it is withholding vital UFO information from the public, thereby preventing proper evaluation. On numerous occasions the Air Force has publicized its conclusions relative to UFOs and explained the evaluation process. This group, nevertheless, continues to claim that UFOs are objects from outer space and demand Congressional hearings on the subject. The continued interest of this small segment is understandable because the subject is so novel and fascinating that it supports over a hundred organizations of one type or another. Most of these organizations publish news releases or magazines and they expect the Air Force to furnish them grist for their publication. Needless to say, the Air Force does not have the resources alloted to this project to fill these numerous individual requests which organizations make for copies of investigative reports and other related material.

In addition, the Air Force would be remiss in its duty to the American public if by its assistance, it encouraged these organizations in their sensational claims and contentions. The Air Force cannot give them individual attention but rather makes periodic releases available to those groups through established news channels. In so doing, it shows no partiality to any person or organization, nor does it place itself in the position of placing its stamp of approval on or giving preferential treatment to any one of them.

This press release approach used by all branches of Government, is considered censorship by many of these organizations and because the Air Force will not favor them with their individual attention they contend that it is keeping vital information from the public. The Air Force was compelled to adopt the press release approach because in the past when it furnished factual information to certain writers of UFO books upon their individual requests, the action was interpreted as granting approval and clearance to the books in which the information was used.

If the Air Force withholds certain information from the public on UFOs, such as the names of individuals reporting sightings, it is not done for the purpose of depriving the public of vital information necessary for proper evaluation. Nor is it done because there is scientific proof of the existence of space craft from other planets and it does not wish to alarm the American public. It is done in the majority of instances to protect the people involved from the idle curiosity of the sensation seekers. In a few instances it is done to keep from compromising investigative processes and if necessary, it is done for legitimate security. This last instance is one this small group fails to recognize as legitimate and yet every citizen must recognize that specific information could benefit a potential enemy. An example of this would be a reported radar UFO sighting by one of our defense radar installations which was caused by some specific weather condition such as a temperature inversion. If this certain set of atmospheric conditions would always, or generally, produce spurious radar returns on the scope, it would become classified information. The reason for this is obvious. Any potential enemy could try to penetrate our defense radar at the particular geographical point when the atmospheric conditions were just right.

In this regard, it should be repeated here that there is no classified material on file in the Air Force which proves or concludes that interplanetary space travel exists in reverse.

The Air Force has a tremendous task in defending this country against weapons systems which we know actually exist and are in the hands of our potential enemies. To divert more men and money away from this most serious mission into a greatly enlarged program for the investigation and defense against UFOs about which we have been unable to discover one iota of tangible scientific evidence, would seriously jeopardize the security of this country against a known proven threat; would be allowing the sensation seekers to dictate our defense policies and would lay the Air Force open to the charge of gross imprudence.

The Air Force does not deny that unknown objects have been seen by responsible people. It is in the interpretation of these sightings that they are questioned. From its investigations covering the past 13 year period, the Air Force contends that when the evidence of these sightings has been sifted through the scientific criteria it has always led to the conclusions that the objects were not space craft and they did not constitute a threat against the security of this country. As an act of faith, the UFO can be considered manned or unmanned craft from outer space. But as a scientific fact there has been no authenticated scientific evidence presented to or discovered by the Air Force to support this conclusion.

Finally, when space travel is accomplished by the United States or should the Air Force discover that spaceships exist from out of this world, official Government announcements would be made immediately through the Department of Defense to this effect.

And that's the story to date. There just are not any manned spaceships yet. There are what might be called "flying saucers" under development but they are really aircraft which operate within the atmosphere and are definitely not spaceships. Present development designs limit these air cars to 10,000 feet altitude and the majority of them are only designed to skim along approximately one foot off the ground on their own self-made cushion of low pressure low velocity air. Companies developing these cars are the Hovercraft Co. of England, Aero, Ltd. of Canada, and Curtis-Wright and Ford Motor Co. of this country.

APPENDICES 1-11

AIR FORCE REGULATION
NO. 200-2

DEPARTMENT OF THE AIR FORCE
Washington, *14 September 1959*

*AFR 200-2
1-2

Intelligence

UNIDENTIFIED FLYING OBJECTS (UFO)

This regulation establishes the responsibility and procedure for reporting information and evidence on unidentified flying objects (UFO) and for releasing pertinent information to the general public.

SECTION A—GENERAL

	Paragraph
Background Information	1
Definitions	2
Objectives	3
Responsibilities	4
Guidance	5
Reporting UFO Information	6

SECTION B—PUBLIC RELATIONS, INFORMATION, CONTACTS, AND RELEASES

Maintaining Public Relations	7
Releasing Information	8
Exceptions	9
Release by Non-Air Force Sources	10
Contacts	11

SECTION C—PREPARING AND SUBMITTING REPORTS

General Information	12
Methods for Transmitting Reports	13
Where To Submit Reports	14
Basic Reporting Data and Format	15
Negative or Inapplicable Data	16
Comments of Preparing Officer	17
Classification	18
Reporting Physical Evidence	19

SECTION A—GENERAL

1. Background Information. The Air Force investigation and analysis of UFO's over the United States are directly related to its responsibility for the defense of the United States. Because prompt reporting and rapid identification are necessary to carry out the second of the four phases of air defense—detection, identification, interception, and destruction, the Air Force maintains the Unidentified Flying Object Program. Successful implementation of the program requires strict compliance with this regulation by all commanders.

2. Definitions. To insure proper and uniform usage in UFO screenings, investigations, and reportings, the objects are defined as follows:

a. *Familiar or Known Objects.* Aircraft, birds, balloons, kites, searchlights, and astronomical bodies (meteors, planets, stars).

b. *Unidentified Aircraft:*

(1) Flying objects determined to be aircraft. These generally appear as a result of ADIZ violations and often prompt the UFO reports submitted by the general public. They are readily identifiable as, or known to be, aircraft, but their type, purpose, origin, and destination are unknown. Air Defense Command is responsible for reports of "unknown" aircraft and they should not be reported as UFO's under this regulation.

(2) Aircraft flares, jet exhausts, condensation trails, blinking or steady lights observed at night, lights circling or near airports and airways, and other similar phenomena known to be emanating from, or to be indications of aircraft. These should not

---
*This regulation supersedes AFRs 200-2, 5 February 1958, and 200-2A, 16 April 1959.

OPI: AFCIN
DISTRIBUTION: S

be reported under this regulation as they do not fall within the definition of a UFO.

(3) Pilotless aircraft and missiles.

c. *Unidentified Flying Objects.* Any airborne object which, by performance, aerodynamic characteristics, or unusual features, does not conform to known aircraft or missiles, or which does not correspond to definitions in a and b above.

3. **Objectives.** Air Force interest in UFO's is three-fold: First, as a possible threat to the security of the United States and its forces; second, to determine the technical or scientific characteristics of any such UFO's; third, to explain or identify all UFO sightings as defined in paragraph 2c.

a. *Air Defense.* The great majority of flying objects reported have been found to be conventional, familiar things of no great threat to the security of the United States and its possessions. However, since the possibility cannot be ignored that UFO's reported may be hostile or new foreign air vehicles of unconventional design, it is imperative that sightings be reported rapidly, factually, and as completely as possible.

b. *Technical and Scientific.* The Air Force will continue to collect and analyze reports of UFO's until all can be scientifically or technically explained or until such time as it is determined that the full potential of a sighting has been exploited. In performance of this task the following factors should be kept in mind:

(1) To measure scientific advances, the Air Force must have the latest experimental and developmental information on new or unique air vehicles or weapons.

(2) The possibility exists that foreign air vehicles of revolutionary configuration or propulsion may be developed.

(3) There is a need for further scientific knowledge in such fields as geophysics, astronomy, and the upper atmosphere which the study and analysis of UFO's and similar aerial phenomena may provide.

(4) The reporting of all pertinent factors will have a direct bearing on scientific analyses and conclusions of UFO sightings.

c. *Reduction of Percentage of UFO "Unidentifieds."* Air Force activities must reduce the percentage of unidentifieds to the minimum. Analysis thus far has provided explanation for all but a few of the sightings reported. These unexplained sightings are carried statistically as unidentifieds. If more immediate, detailed objective data on the unknowns had been available, probably these too could have been explained. However, due to the human factors involved, and the fact that analyses of UFO sightings are based primarily on the personal impressions and interpretations of the observers, rather than on accurate scientific data or facts obtained under controlled conditions, it is improbable that all of the unidentifieds can be eliminated.

4. **Responsibilities:**

a. *Reporting.* Base commanders will report all information and evidence of UFO sightings, including information and evidence received from other services, Government agencies, and civilian sources. Investigators will be authorized to make telephone calls from the investigation area direct to the Air Technical Intelligence Center (ATIC), Wright-Patterson Air Force Base, Ohio (CLearwater 3-7111, ext. 69216). The purpose of the calls is to report high priority findings. (See section C.)

b. *Investigation.* The commander of the Air Force base nearest the location of the reported UFO sighting will conduct all investigative action necessary to submit a complete initial report of a UFO sighting. Every effort will be made to resolve the sighting in the initial investigation. A UFO sighting reported to an Air Force base other than that closest to the scene of such sighting will be referred immediately to the commander of the nearest Air Force base for appropriate action. (See paragraph 6.)

c. *Analysis.* The ATIC will analyze and evaluate:

(1) Information and evidence reported within the United States after the investigators of the responsible Air Force base nearest the sighting have exhausted their efforts to identify the UFO.

(2) Information and evidence collected in oversea areas.

Note. Exceptions: The ATIC, independently or in participation with pertinent Air Force activities, may conduct any additional investigations necessary to further or conclude its analyses or findings.

d. *Public Relations and Information Services.* The Office of Information Services, Office of the Secretary of the Air Force, will be responsible for releasing information on sightings, and, in coordination with ATIC, for answering correspondence from the public regarding UFO's. (See paragraphs 7 and 8.)

e. *Congressional Inquiries.* The Office of Legislative Liaison will:
   (1) In coordination with the ATIC and/or the Office of Information Services, when necessary, answer all congressional mail regarding UFO's addressed to the Secretary of the Air Force and Headquarters USAF.
   (2) Forward those inquiries which are scientific and technical to the ATIC for information on which to base a reply. The ATIC will return this information to the Office of Legislative Liaison for reply to the inquiry.
   (3) Process requests from congressional sources in accordance with AFR 11-7.

f. *Cooperation.* All Air Force activities will cooperate with Air Force UFO investigators to insure the economical and prompt success of investigations and analyses. When feasible, this cooperation will include furnishing air or ground transportation and other assistance.

5. Guidance. The thoroughness and quality of a report or investigation of UFO's are limited only by the skill and resourcefulness of the person who receives the initial information and/or prepares the report. The usefulness and value of any report or investigation depend on the accuracy and timeliness of its contents. Following are aids for screening, evaluating, and reporting sightings:

a. Careful study of the logic, consistency, and coherence of the observer's report. An interview with the observer by personnel preparing the report is especially valuable in determining the source's reliability and the validity of the information given. Particular attention should be given to the observer's age, occupation, and education and whether his occupation involves observation reporting or technical knowledge. When reporting that a witness is completely familiar with certain aspects of a sighting, his or her specific qualifications should be indicated.

b. Theodolite measurements of changes of azimuth, and elevation and angular size.

c. Interception, identification, or air search if appropriate and within the scope of air defense regulations.

d. When feasible, contact with local aircraft control and warning (ACW) units, pilots and crews of aircraft aloft at the time and place of sighting. Also, contact with any other persons or organizations that may have factual data on the UFO or can offer corroborating evidence—visual, electronic, or other.

e. Consultation with military or civilian weather forecasters for data on tracks of weather balloons released in the area and any unusual meteorological activity which may have a bearing on the UFO.

f. Consultation with navigators and astronomers in the area to determine whether any astronomical body or phenomenon would account for the sighting.

g. Contact with military and civilian tower operators, air operations units, and airlines to determine whether the sighting could have been an aircraft. Local units of the Federal Aviation Agency (FAA) are often of assistance in this regard.

h. Contact with persons who may know of experimental aircraft of unusual configuration, rocket and guided missile firings, or aerial tests in the area.

i. Contact with photographic units or laboratories. Usually, these installations have several cameras available for specialized intelligence or investigative work. Photography is an invaluable tool and, where possible, should be used in investigating and analyzing UFO sightings. (See paragraph 19.)

j. Whenever possible, selecting as a UFO sighting investigator an individual with a scientific or technical background as well as experience as an investigator.

6. Reporting UFO Information. Both the Assistant Chief of Staff Intelligence, Headquarters USAF, and the Air Defense Command have a direct and immediate interest in the facts pertaining to UFO's reported within the United States.

a. All Air Force activities will conduct UFO investigations to the extent necessary for their required reporting action (see paragraphs 15, 16, and 17). However, investigations should not be carried beyond this point, unless such action is directed by Assistant Chief of Staff, Intelligence, Headquarters USAF, or the preparing officer believes the magnitude (intelligence significance or public relations) of the case warrants full scale investigation. Telephone contact should be made with the ATIC (CLearwater 3-7111, ext. 69216) at Wright-Patterson Air Force Base, Ohio, to obtain verbal authority for continued investigation. This should be so noted in the preliminary report. (Foreign activities will proceed on their own judgment and so advise the ATIC in the preliminary message.)

b. After initial reports are submitted, the ATIC may require additional data, such as narrative statements, sketches, marked maps

AFR 200-2
6-12

and charts, and other required data, which can be supplied more quickly and economically by the Air Force activity that made the initial report. Therefore, ATIC is authorized to contact the appropriate Air Force activity.

c. Direct communication is authorized between ATIC and other Air Force activities in matters pertaining to UFO investigations. Specifically, the ATIC may call upon the Commander, 1137th Field Activities Group, Fort Belvoir, Virginia, to conduct further field investigation if review of the initial report indicates such a requirement. In this event, the AISS investigating will prepare the final report. (See paragraph 4b.)

SECTION B—PUBLIC RELATIONS, INFORMATION, CONTACTS AND RELEASES

7. Maintaining Public Relations. The Office of Information Services is responsible for:

a. In coordination with the ATIC when necessary, maintaining contact with the public and the press on all aspects of the UFO program and its related activities.

b. Releasing information on UFO sightings and results of investigations.

c. Periodically releasing information on this subject to the general public.

d. Processing, answering, and taking action on correspondence received from the general public, pertaining to the public relations, interest, and informational aspects of the subject. (See paragraph 9.) This office will forward correspondence and queries which are purely technical and scientific to ATIC for information on which to base a reply.

8. Releasing Information. All information or releases concerning UFO's, regardless of origin or nature, will be released to the public or unofficial persons or organizations by the Office of Information Services, Office of the Secretary of the Air Force. This includes replies to correspondence (except congressional inquiries) submitted direct to ATIC, and other Air Force activities by private individuals requesting comments or results or analysis and investigations of sightings.

9. Exceptions. In response to local inquiries resulting from any UFO reported in the vicinity of an Air Force base, information regarding a sighting may be released to the press or the general public by the commander of the Air Force base concerned only if it has been *positively identified as a familiar or known object*. Care should be exercised not to reveal any classified aspects of the sighting or names of persons making reports. (See paragraph 18.) If the sighting is unexplainable or difficult to identify, because of insufficient information or inconsistencies, the only statement to be released is the fact that the sighting is being investigated and information regarding it will be released at a later date. If investigative action has been completed, the fact that the results of the investigation will be submitted to the ATIC for review and analysis may be released. Further inquiries should be referred to the local Office of Information Services.

. 10. Release by Non-Air Force Sources. If newsmen, writers, publishers, or private individuals desire to release unofficial information concerning a UFO sighting, every effort will be made to assure that the statements, theories, opinions, and allegations of these individuals or groups will not be associated with or represented as being official information.

11. Contacts. Private individuals or organizations requesting Air Force interviews, briefings, lectures, or private discussions on UFO's will be referred to the Office of Information Services, Office of the Secretary of the Air Force. Air Force personnel, other than those of the Office of Information Services, will not contact private individuals on UFO cases nor will they discuss their operations and functions with unauthorized persons unless so directed, and then only on a "need-to-know" basis.

SECTION C—PREPARING AND SUBMITTING REPORTS

12. General Information:

a. Paragraphs 2 and 5 will be used as an aid and guidance to screenings, investigations, and reportings. The format will be as outlined in paragraph 15. Activities initially receiving reports of aerial objects and phenomena will screen the information to determine if the report concerns a valid UFO within the definition of paragraph 2c. Reports not within that definition will not be considered for further action under the provisions of this regulation.

b. To assist activities and personnel responsible for handling, screening, and processing initial, incoming UFO information, the general sources and types of reports are given here:

    (1) Generally, initial UFO reports are received from two sources:

        (a) Civilian (airline, private and professional pilots, tower operators, technical personnel, casual observ-

ers, and the public in general), by correspondence, telephone, or personal interview;
(b) Military units and personnel (pilots, observers, radar operators, aircraft control and warning units, etc.), by telephone, electrical message, or personal interview;

(2) Generally, UFO reports received from civilian sources are of two types:
(a) Those referring strictly to an observed UFO, containing either detailed or meager information;
(b) Those referring only in part to an observed UFO, but primarily requesting information on some aspect of the UFO program.

c. Reports considered to fall primarily in a public relations or information service category (see paragraphs 4d, 7, 8, 9, and b(2) above) should be referred to the Office of Information Services. UFO data sufficient for investigation and/or analysis may be extracted before referral to that office.

13. Methods for Transmitting Reports:

a. Together with any necessary screenings and investigations that must be performed preparatory to reporting, all information on UFO's will be reported promptly. Reports under 3 days from date of sighting will be electrically transmitted with a "Priority" precedence. Electrically transmitted reports over 3 days old should carry a "Routine" precedence.

b. Written reports of sightings over 3 days old may be submitted on AF Form 112, Air Intelligence Information Report (AIIR) and AF Form 112A, supplement to AF Form 112 (see paragraphs 14 and 15); however, their use should be kept to a minimum in reporting initial sightings. The delays often involved in processing and transmitting AF Forms 112 through channels may make followup investigations difficult, producing only limited usable information. This factor must be considered in cases where an immediate investigation or study of a reported sighting is considered necessary. Reporting by electrical means will eliminate delays. If requested by ATIC, a followup and/or complete report of all sightings initially reported electrically will be submitted on AF Form 112.

14. Where To Submit Reports:

a. *Electrical Reports.* Submit multiple addressed electrical reports to:
(1) Air Defense Command, Ent AFB, Colorado
(2) Nearest Air Division (Defense). (For United States only.)
(3) Air Technical Intelligence Center, Wright-Patterson AFB, Ohio
(4) HQ USAF (AFCIN), Wash. 25, D.C.
(5) Secretary of the Air Force (SAFIS), Wash. 25, D.C.

b. *Written Reports:* (Basic letters and AF Forms 112.)
(1) Within the United States, submit all reports direct to ATIC. ATIC will reproduce each report and distribute it to interested intelligence activities in the United States and to Office of Information Services, if such action is considered necessary.
(2) Outside the United States, submit reports as prescribed in "Intelligence Collection Instruction" (ICI) June 1954, direct to:
Hq USAF (AFCIN) Wash 25, D.C.

c. *Reports from Civilians.* Where possible, civilian sources contemplating reporting UFO's should be advised to submit the report, for processing and transmission, to the nearest Air Force base, other than ATIC.

15. Basic Reporting Data and Format.
Show the abbreviation "UFO" at the beginning of the text of all electrical reports and in the subject of written reports. Include in all reports the data required, in the order shown below:

a. *Description of the Object(s):*
(1) Shape.
(2) Size compared to a known object (use one of the following terms: Head of a pin, pea, dime, nickel, quarter, half dollar, silver dollar, baseball, grapefruit, or basketball) held in the hand at about arm's length.
(3) Color.
(4) Number.
(5) Formation, if more than one.
(6) Any discernible features or details.
(7) Tail, trail, or exhaust, including size of same compared to size of object(s).
(8) Sound. If heard, describe sound.
(9) Other pertinent or unusual features.

b. *Description of Course of Object(s):*
(1) What first called the attention of observer(s) to the object(s)?
(2) Angle or elevation and azimuth of objects(s) when first observed.

(3) Angle or elevation and azimuth of object(s) upon disappearance.
(4) Description of flight path and maneuvers of object(s).
(5) How did the object(s) disappear? (Instantaneously to the North, etc.)
(6) How long was the object(s) visible? (Be specific, 5 minutes, 1 hour, etc.)

c. *Manner of Observation:*
(1) Use one or any combination of the following items: Ground-visual, ground-electronic, air electronic. (If electronic, specify type of radar.)
(2) Statement as to optical aids (telescopes, binoculars, etc.) used and description thereof.
(3) If the sighting is made while airborne, give type of aircraft, identification number, altitude, heading, speed, and home station.

d. *Time and Date of Sighting:*
(1) Zulu time-date group of sighting.
(2) Light conditions. (Use one of the following terms: Night, day, dawn, dusk.)

e. *Location of Observer(s).* Exact latitude and longitude of each observer, and/or geographical position. A position with reference to a known landmark also should be given in electrical reports, such as "2mi N of Deeville;" "3mi SW of Blue Lake." Typographical errors or "garbing" often result in electrically transmitted messages, making location plots difficult or impossible.

Example: 89 45N, 192 71W for 39 45N, 102 21W.

f. *Identifying Information on Observer(s):*
(1) Civilian—Name, age, mailing address, occupation, and estimate of reliability.
(2) Military—Name, grade, organization, duty, and estimate of reliability.

g. *Weather and Winds—Aloft Conditions at Time and Place of Sightings:*
(1) Observer(s) account of weather conditions.
(2) Report from nearest AWS or U.S. Weather Bureau Office of wind direction and velocity in degrees and knots at surface, 6,000', 10,000', 16,000', 20,000', 30,000', 50,000', and 80,000' if available.
(3) Ceiling.
(4) Visibility.

(5) Amount of cloud cover.
(6) Thunderstorms in area and quadrant in which located.
(7) Temperature gradient.

h. Any other unusual activity or condition, meteorological, astronomical, or otherwise, which might account for the sighting.

i. Interception or identification action taken (such action may be taken whenever feasible, complying with existing air defense directives).

j. Location, approximate altitude, and general direction of flight of any air traffic or balloon releases in the area which could possibly account for the sighting.

k. Position title and comments of the preparing officer, including his preliminary analysis of the possible cause of the sighting(s). (See paragraph 17.)

l. Existence of physical evidence, such as materials and photographs.

16. Negative or Inapplicable Data. Even though the source does not provide or has not been asked for specific information by an interviewer, do not use the words "negative" or "unidentified" until all logical leads to obtain the information outlined under paragraph 15 have been exhausted. For example, information on weather conditions in the area, as requested in paragraph 15g may be obtained from the local military or civilian weather facility. Use the phrase "not applicable" (N/A) only when the question does not apply to the particular sighting being investigated.

17. Comments of Preparing Officer. The preparing officer will make a preliminary analysis and a comment on the possible cause or identity of the object being reported, together with a statement supporting his comment and analysis. Every effort will be made to obtain pertinent items of information and to test all possible leads, clues, and hypotheses concerning the identity or explanation of the sighting. (See paragraph 5.) The preparing officer receiving the report initially is in a much better position to conduct an "on-the-spot" survey or followup than subsequent investigative personnel and analysts who may be far removed from the area, and who may arrive too late to obtain vital data or the missing information necessary for firm conclusions.

18. Classification. Do not classify reports unless data requested in paragraph 15 require classification. Classify reports primarily to protect:

a. Names of sources reporting UFO's and other principals involved, if so requested by these persons or considered necessary;

b. Intelligence, investigative, intercept, or analytical methods or procedures;

c. Location of radar and other classified sites, units, and equipment;

d. Information on certain types, characteristics, and capabilities of classified aircraft, missiles, or devices that may be involved in the sighting.

19. Reporting Physical Evidence. Report promptly the existence of physical evidence (photographic or material). All physical evidence forwarded to the ATIC should be marked for the attention of AFCIN-4E4g.

a. *Photographic:*

(1) *Still Photographs.* Forward the negative and two prints. Title the prints and the negatives, or indicate the place, time, and date of the incident.

(2) *Motion Pictures.* Obtain the original film. Examine the film strip for apparent cuts, alterations, obliterations, or defects. In the report comment on any irregularities, particularly if received from other than official sources.

(3) *Supplemental Photographic Information.* Negatives and prints often are insufficient to provide certain valid data or to permit firm conclusions. (See AFM 200-9—a classified document receiving limited distribution.) Information that will aid in plotting or in estimating distances, apparent size and nature of object, probable velocity, and movements includes:

(a) Type and make of camera,
(b) Type, focal length, and make of lens,
(c) Brand and type of film,
(d) Shutter speed used,
(e) Lens opening used, that is, "f" stop,
(f) Filters used,
(g) Was tripod or solid stand used,
(h) Was "panning" used,
(i) Exact direction camera was pointing with relation to true north, and its angle with respect to the ground.

(4) *Other Camera Data.* If supplemental information cannot be obtained, the minimum camera data required are the type of camera, and the smallest and largest "f" stop and shutter-speed readings of the camera.

(5) *Radar.* Forward two copies of each still-camera photographic print. Title radarscope photographic prints in accordance with AFR 95-7. Classify radarscope photographs in accordance with section XII, AFR 205-1, 1 April 1959.

Note: If possible, develop photographic film before forwarding. If undeveloped film is forwarded, mark it conspicuously to indicate this fact. Undeveloped film often has been destroyed by exposure during examinations made while en route through mail channels to final addressees.

b. *Material.* Each Air Force echelon receiving suspected or actual UFO material will safeguard it in a manner to prevent any defacing or alterations which might reduce its value for intelligence examination and analysis.

c. *Photographs, Motion Pictures, and Negatives Submitted by Individuals.* Individuals often submit photographic and motion picture material as part of their UFO reports. All original material submitted, will be returned to the individual upon completion of necessary studies, analyses, and duplication by the Air Force.

BY ORDER OF THE SECRETARY OF THE AIR FORCE:

OFFICIAL:

J. L. TARR
Colonel, USAF
Director of Administrative Services

THOMAS D. WHITE
Chief of Staff

CHANGE	AFR 200-2A
	6c

AIR FORCE REGULATION	DEPARTMENT OF THE AIR FORCE
NO. 200-2A	Washington, *2 February 1960*

Intelligence

UNIDENTIFIED FLYING OBJECTS (UFO)

AFR 200-2, 14 September 1959, is changed as follows:

6c. Direct communication is authorized between ATIC and other Air Force activities in matters pertaining to UFO investigation. Specifically, the ATIC may call upon the Commander, 1127th Field Activity Group, Fort Belvoir, Virginia, to conduct further field investigations if review of the initial report indicates such a requirement. In this event, the Headquarters 1127th USAF Field Activity Group will prepare the final report.

BY ORDER OF THE SECRETARY OF THE AIR FORCE:

OFFICIAL:

THOMAS D. WHITE
Chief of Staff

J. L. TARR
Colonel, USAF
Director of Administrative Services

DISTRIBUTION: S

U. S. AIR FORCE TECHNICAL INFORMATION SHEET

This questionnaire has been prepared so that you can give the U. S. Air Force as much information as possible concerning the unidentified aerial phenomenon that you have observed. Please try to answer as many questions as you possibly can. The information that you give will be used for research purposes, and will be regarded as confidential material. Your name will not be used in connection with any statements, conclusions, or publications without your permission. We request this personal information so that, if it is deemed necessary, we may contact you for further details.

1. When did you see the object?

    _____   _____   _____
      Day   Month   Year

2. Time of day: _____   _____
                  Hour   Minutes

    (Circle One):   A.M.   or   P.M.

3. Time zone:
    (Circle One):   a. Eastern
                    b. Central
                    c. Mountain
                    d. Pacific
                    e. Other _____

    (Circle One):   a. Daylight Saving
                    b. Standard

4. Where were you when you saw the object?

    _____   _____   _____
       Nearest Postal Address   City or Town   State or Country
    Additional remarks: _____

5. Estimate how long you saw the object. _____   _____   _____
                                            Hours   Minutes   Seconds

    5.1 Circle one of the following to indicate how certain you are of your answer to Question 5.

        a. Certain   c. Not very sure
        b. Fairly certain   d. Just a guess

6. What was the condition of the sky?

    (Circle One):   a. Bright daylight   d. Just a trace of daylight
                    b. Dull daylight   e. No trace of daylight
                    c. Bright twilight   f. Don't remember

7. IF you saw the object during DAYLIGHT, TWILIGHT, or DAWN, where was the SUN located as you looked at the object?

    (Circle One):   a. In front of you   d. To your left
                    b. In back of you   e. Overhead
                    c. To your right   f. Don't remember

ATIC FORM NO. 164 (13 OCT 54)

8. IF you saw the object at NIGHT, TWILIGHT, or DAWN, what did you notice concerning the STARS and MOON?

   8.1 STARS (Circle One):

   a. None
   b. A few
   c. Many
   d. Don't remember

   8.2 MOON (Circle One):

   a. Bright moonlight
   b. Dull moonlight
   c. No moonlight — pitch dark
   d. Don't remember

9. Was the object brighter than the background of the sky?

   (Circle One):   a. Yes   b. No   c. Don't remember

10. IF it was BRIGHTER THAN the sky background, was the brightness like that of an automobile headlight?:

    (Circle One) a. A mile or more away (a distant car)?
    b. Several blocks away?
    c. A block away?
    d. Several yards away?
    e. Other

11. Did the object: (Circle One for each question)

a. Appear to stand still at any time?	Yes	No	Don't Know
b. Suddenly speed up and rush away at any time?	Yes	No	Don't Know
c. Break up into parts or explode?	Yes	No	Don't Know
d. Give off smoke?	Yes	No	Don't Know
e. Change brightness?	Yes	No	Don't Know
f. Change shape?	Yes	No	Don't Know
g. Flicker, throb, or pulsate?	Yes	No	Don't Know

12. Did the object move behind something at anytime, particularly a cloud?

    (Circle One):   Yes   No   Don't Know.   IF you answered YES, then tell what it moved behind: _____

13. Did the object move in front of something at anytime, particularly a cloud?

    (Circle One):   Yes   No   Don't Know.   IF you answered YES, then tell what it moved in front of: _____

14. Did the object appear: (Circle One):   a. Solid?   b. Transparent?   c. Don't Know.

15. Did you observe the object through any of the following?

a. Eyeglasses	Yes	No	e. Binoculars	Yes	No
b. Sun glasses	Yes	No	f. Telescope	Yes	No
c. Windshield	Yes	No	g. Theodolite	Yes	No
d. Window glass	Yes	No	h. Other _____		

16. Tell in a few words the following things about the object.

   a. Sound _____

   b. Color _____

17. Draw a picture that will show the shape of the object or objects. Label and include in your sketch any details of the object that you saw such as wings, protrusions, etc., and especially exhaust trails or vapor trails. Place an arrow beside the drawing to show the direction the object was moving.

18. The edges of the object were:

   (Circle One): a. Fuzzy or blurred   e. Other _____
                  b. Like a bright star   _____
                  c. Sharply outlined   _____
                  d. Don't remember

19. IF there was MORE THAN ONE object, then how many were there? _____
    Draw a picture of how they were arranged, and put an arrow to show the direction that they were traveling.

20. Draw a picture that will show the motion that the object or objects made. Place an "A" at the beginning of the path, a "B" at the end of the path, and show any changes in direction during the course.

21. IF POSSIBLE, try to guess or estimate what the real size of the object was in its longest dimension.
    _____ feet.

22. How large did the object or objects appear as compared with one of the following objects *held in the hand* and at about arm's length?

    (Circle One):
    a. Head of a pin
    b. Pea
    c. Dime
    d. Nickel
    e. Quarter
    f. Half dollar
    g. Silver dollar
    h. Baseball
    i. Grapefruit
    j. Basketball
    k. Other _____

22.1 (Circle One of the following to indicate how certain you are of your answer to Question 22.

    a. Certain
    b. Fairly certain
    c. Not very sure
    d. Uncertain

23. How did the object or objects disappear from view? _____
_____
_____

24. In order that you can give as clear a picture as possible of what you saw, we would like for you to imagine that you could construct the object that you saw. Of what type material would you make it? How large would it be, and what shape would it have? Describe in your own words a common object or objects which when placed up in the sky would give the same appearance as the object which you saw.

25. Where were you located when you saw the object? (Circle One):

   a. Inside a building
   b. In a car
   c. Outdoors
   d. In an airplane
   e. At sea
   f. Other _____

26. Were you (Circle One)

   a. In the business section of a city?
   b. In the residential section of a city?
   c. In open countryside?
   d. Flying near an airfield?
   e. Flying over a city?
   f. Flying over open country?
   g. Other _____

27. What were you doing at the time you saw the object, and how did you happen to notice it?

   _____
   _____
   _____

28. IF you were MOVING IN AN AUTOMOBILE or other vehicle at the time, then complete the following questions:

   28.1 What direction were you moving? (Circle One)

a. North	c. East	e. South	g. West
b. Northeast	d. Southeast	f. Southwest	h. Northwest

   28.2 How fast were you moving? _____ miles per hour.

   28.3 Did you stop at any time while you were looking at the object?
       (Circle One)   Yes   No

29. What direction were you looking when you first saw the object? (Circle One)

a. North	c. East	e. South	g. West
b. Northeast	d. Southeast	f. Southwest	h. Northwest

30. What direction were you looking when you last saw the object? (Circle One)

a. North	c. East	e. South	g. West
b. Northeast	d. Southeast	f. Southwest	h. Northwest

31. If you are familiar with bearing terms (angular direction), try to estimate the number of degrees the object was from true North and also the number of degrees it was upward from the horizon (elevation).

   31.1 When it first appeared:
       a. From true North _____ degrees.
       b. From horizon _____ degrees.

   31.2 When it disappeared:
       a. From true North _____ degrees.
       b. From horizon _____ degrees.

32. In the following sketch, imagine that you are at the point shown. Place an "A" on the curved line to show how high the object was above the horizon (skyline) when you *first* saw it. Place a "B" on the same curved line to show how high the object was above the horizon (skyline) when you *last* saw it.

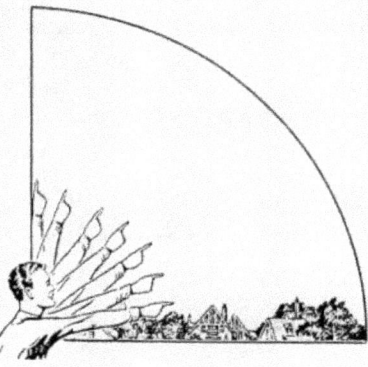

33. In the following larger sketch place an "A" at the position the object was when you *first* saw it, and a "B" at its position when you *last* saw it. Refer to smaller sketch as an example of how to complete the larger sketch.

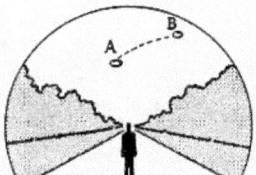

34. What were the weather conditions at the time you saw the object?

    34.1 CLOUDS *(Circle One)*

        a. Clear sky
        b. Hazy
        c. Scattered clouds
        d. Thick or heavy clouds
        e. Don't remember

    34.2 WIND *(Circle One)*

        a. No wind
        b. Slight breeze
        c. Strong wind
        d. Don't remember

    34.3 WEATHER *(Circle One)*

        a. Dry
        b. Fog, mist, or light rain
        c. Moderate or heavy rain
        d. Snow
        e. Don't remember

    34.4 TEMPERATURE *(Circle One)*

        a. Cold
        b. Cool
        c. Warm
        d. Hot
        e. Don't remember

35. When did you report to some official that you had seen the object?

    _____   _____   _____
        Day   Month   Year

36. Was anyone else with you at the time you saw the object?

    (Circle One)   Yes   No

    36.1 IF you answered YES, did they see the object too?

    (Circle One)   Yes   No

    36.2 Please list their names and addresses:

37. Was this the first time that you had seen an object or objects like this?

    (Circle One)   Yes   No

    37.1 IF you answered NO, then when, where, and under what circumstances did you see other ones?

    _____
    _____
    _____
    _____

38. In your opinion what do you think the object was and what might have caused it?

39. Do you think you can estimate the speed of the object?

    (Circle One)   Yes   No

IF you answered YES, then what speed would you estimate? _____ m.p.h.

40. Do you think you can estimate how far away from you the object was?

    (Circle One)   Yes   No

IF you answered YES, then how far away would you say it was? _____ feet.

41. Please give the following information about yourself:

NAME _____ _____ _____
              Last Name   First Name   Middle Name

ADDRESS _____ _____ _____ _____
                Street   City   Zone   State

TELEPHONE NUMBER _____

What is your present job? _____

Age _____   Sex _____

Please indicate any special educational training that you have had.

    a. Grade school _____   e. Technical school _____
    b. High school _____   (Type) _____
    c. College _____   f. Other special training _____
    d. Post graduate _____

42. Date you completed this questionnaire: _____ _____ _____
                                                               Day   Month   Year

U. S. AIR FORCE TECHNICAL INFORMATION SHEET
(SUMMARY DATA)

In order that your information may be filed and coded as accurately as possible, please use the following space to write out a short description of the event that you observed. You may repeat information that you have already given in the questionnaire, and add any further comments, statements, or sketches that you believe are important. Try to present the details of the observation in the order in which they occurred. Additional pages of the same size paper may be attached if they are needed.

NAME _____
(Please Print)

SIGNATURE _____

DATE _____

(Do Not Write in This Space)

CODE:

UFO OBSERVERS INSTRUCTION SHEET
(Sky Diagram)

1. GENERAL:

   a. The diagram represents all of the sky normally visible to the observer, who is pictured standing under the center of the "dome" of the sky. It is designed to show a three-dimensional view of the area centered around the observer at the time of the UFO sighting.

   b. The position of any object in the sky can be described by giving its elevation, or angle upward from the horizon, and its bearing or angle along the horizon, eastward from north.

   (1) Illustrations:

   (a) Elevation is 0 degrees for an object on the horizon, and 90 degrees for the point directly over the observer (zenith). Thus, an object half-way up from the horizon to the zenith has an elevation of 45 degrees.

   (b) Bearing (or "azimuth") is the angle along the horizon, starting from north and moving clockwise eastward. Thus, an object directly toward the east, no matter what its elevation is above the horizon, has a bearing of 90 degrees, an object in the south has a bearing of 180 degrees; toward the west, 270 degrees and so on. North is, of course, zero.

   EXAMPLE: An object is seen in the northeast and one-third way up from horizon to overhead. Thus, the object has a bearing of 45 degrees, and elevation of 30 degrees. Similarly, an object having a bearing of 180 degrees and an elevation of 60 degrees would be seen directly south and two-thirds of the way up from the horizon.

2. PLOTTING THE COURSE OF AN OBJECT ON THE SKY DIAGRAM:

   a. The path of an object across the sky can be shown completely on this diagram simply by connecting with a curved or straight line the various positions the object successively occupies (see example sheet). To aid visualization, the path on the western side of the sky is represented by broken lines; the eastern side in solid lines. Direction of the object is indicated by arrows. The duration of the sighting can be shown by indicating the time at the position, where the object was first and last observed. Where possible, the time at various intermediate positions occupied by the object should also be shown.

   b. The diagram can be made a more effective investigative and analytical tool by making the lines (showing the path of the object) thicker or thinner to indicate any varying brightness of the object observed. This is especially valuable when the object appeared only as a moving light at night. Thus, if a light becomes brighter and then gradually fades, it can be represented by a line becoming increasingly thicker and then gradually thinning out to nothing.

   c. Use of colored pencils is especially recommended if the object changes color or hue during the sighting.

ATIC FORM 164a
(25 July 56)

3. EXAMPLE OF DIAGRAM USE:

   a. _Verbal Description of Example Sighting:_ Object was first sighted in the southeast, about half-way up from the horizon to overhead, at 10:45 PM local time. Its shape or outline was hazy, but appeared round and about the size of a pea (at arm's length) from where observed. It was dim at first but brightened considerably as it got higher in the sky. Its color at this point was bluish white. After about two minutes it crossed to the western part of the sky a little to the north of overhead (zenith) and continued its flight toward the west. At this point its color appeared yellowish white. The light went dim when it got two-thirds of the way to the horizon. It then stopped and hovered for about one minute and then climbed rapidly, going toward the southwest and getting brighter. In less than thirty seconds, it had climbed to an elevation of approximately 60 degrees, and then the light went out abruptly.

   b. _Pictorial Description of the Sighting:_ By referring to the example sheet, notice how simply the above sighting can be portrayed and described, without words, on the example diagram attached here. Note the starting point at bearing 135 degrees (southeast) and elevation 45 degrees (half-way up from the horizon) at 10:45 PM (military time, 2245), and the arrow marking direction of flight. Note also the varying thickness of the line to denote changes in brightness, and the use of the dotted line to indicate its path in the western part of the sky. The "time indications" along the path - 2 minutes to get to the meridian (the north-south overhead line), the hovering for 1 minute, and the ascent in 30 seconds to its complete disappearance, are all shown with a few lines. Thus, the entire sighting can be represented easily on one diagram.

4. FURTHER INSTRUCTIONS AND INFORMATION:

   a. Relatively complex trajectories can easily be shown on a diagram of this type. A number of objects sighted can also be indicated, as can any changing formation. The apparent size and shape of the object should be drawn in, preferably by the observer. In the case of an object changing shape or color, this likewise can be drawn in. As previously pointed out, the use of colored pencils to indicate change of color is very desirable.

   b. The landscaping in the sky diagram is placed there to help visualization. If any prominent landmarks such as known mountains, buildings, water towers, or specific installations, trees, etc., are part of the sighting area, they should be incorporated into the drawing. These landmarks may later prove to be invaluable as location, plotting or reference points.

   c. If you are familiar with the constellations or other heavenly bodies, indicate if possible, the relationship (and movements) of the object with respect to these bodies. This can be sketched on either page 6, item 33 or pages 9-10 of "Summary Data" sheet. Typical examples that can be easily illustrated: "...The object seemed to pass very slowly between the two bottom stars on the handle of the Big Dipper, which was in a vertical position, with the handle pointing down," or "...Object was about the size of a tennis ball -- and remained slightly below and about 15 degrees to the left of the moon."

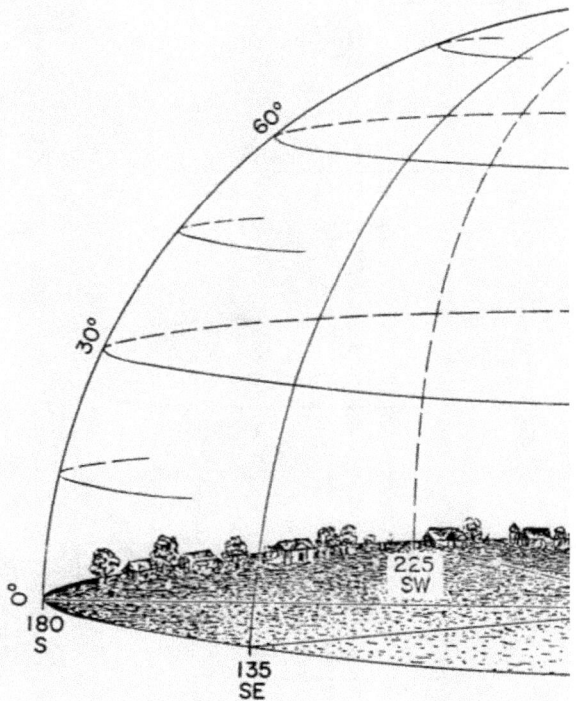

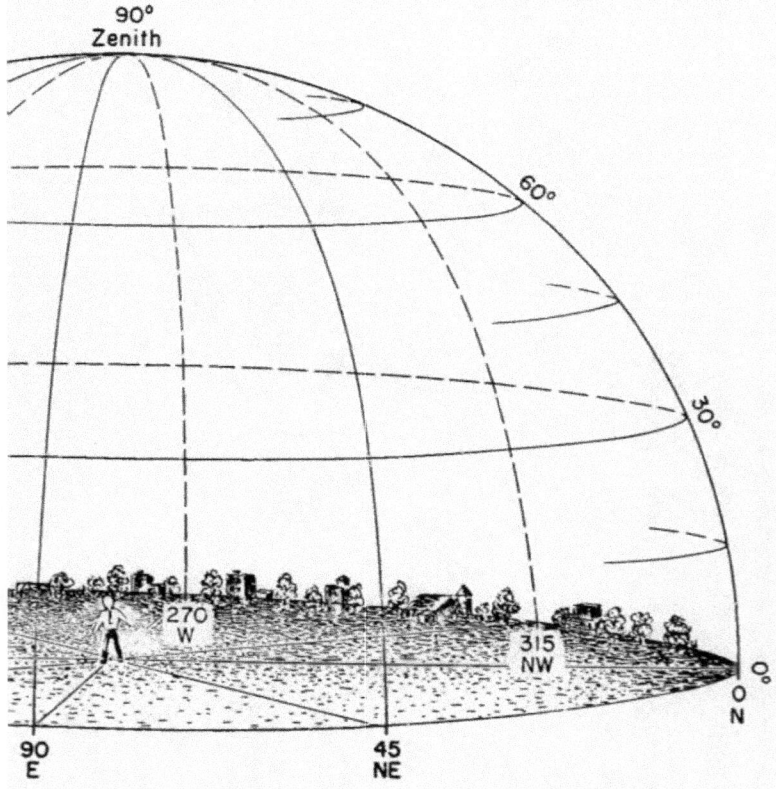

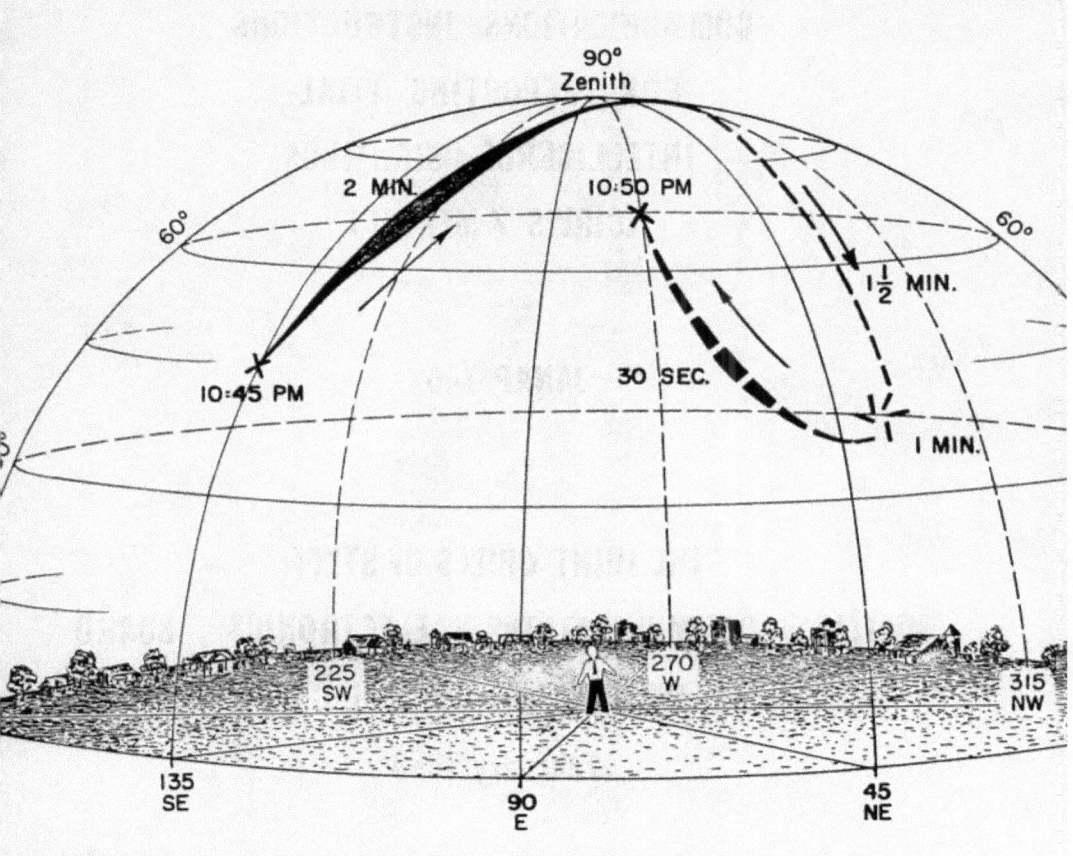

(EXAMPLE SHEET)

JANAP 146(D)

CANADIAN - UNITED STATES COMMUNICATIONS INSTRUCTIONS FOR REPORTING VITAL INTELLIGENCE SIGHTINGS (CIRVIS / MERINT)

JANAP 146

THE JOINT CHIEFS OF STAFF
MILITARY COMMUNICATIONS - ELECTRONICS BOARD
WASHINGTON 25, D.C.
February 1959

ORIGINAL

(Reverse Blank)

JANAP 146(D)

MILITARY COMMUNICATIONS-ELECTRONICS BOARD
WASHINGTON 25, D.C.

1 February 1959

LETTER OF PROMULGATION TO:

The Department of the Army
The Department of the Navy
The Department of the Air Force

Subject: JANAP 146(D)

1. JANAP 146(D), CANADIAN-UNITED STATES COMMUNICATIONS INSTRUCTIONS FOR REPORTING VITAL INTELLIGENCE SIGHTINGS, is an unclassified non-registered publication, prepared by the US Military Communications-Electronics Board in conjunction with the Canadian JCEC(W), for Joint and Canadian use.

2. JANAP 146(D) supersedes JANAP 146(C), and is effective upon receipt for the U.S. JANAP 146(D) will become effective for the Canadian Forces when directed by the appropriate implementing agency.

3. This publication contains military information and is for official use only.

4. Copies and extracts may be made from this publication in the preparation of official publications.

5. Comments and recommendations concerning this publication should not be addressed to the Military Communications-Electronics Board, but to one of the following, as appropriate:

   a. Chief Signal Officer, U.S. Army.
   b. Chief of Naval Operations (DNC), U.S. Navy.
   c. Director of Communications-Electronics, U.S. Air Force.

FOR THE CHAIRMAN, MILITARY COMMUNICATIONS-ELECTRONICS BOARD:

JOSEPH BUSH
Colonel, USAF

E. H. FARRELL
Commander, USN

Secretaries

III   ORIGINAL
                                              (Reverse Blank)

JANAP 146(D)

LIST OF EFFECTIVE PAGES

Subject Matter	Page Numbers	Change in Effect
Title Page	I (Reverse Blank)	Original
Letter of Promulgation to JANAP 146(D) dated 1 February 1959	III (Reverse Blank)	Original
List of Effective Pages	V (Reverse Blank)	Original
Record of Changes	VII (Reverse Blank)	Original
Table of Contents	IX , X	Original
Text		
Chapter 1	1-1 (Reverse Blank)	Original
Chapter 2	2-1 to 2-9 (Reverse Blank)	Original
Chapter 3	3-1 to 3-8	Original

ORIGINAL
(Reverse Blank)

JANAP 146(D)

RECORD OF CHANGES

Identification of Change or Correction and date of same	Date Entered	By whom entered (Signature; rank, grade or rate; name of command)

VII ORIGINAL
(Reverse Blank)

JANAP 146(D)

GENERAL DESCRIPTION AND PURPOSE OF COMMUNICATION
INSTRUCTIONS FOR REPORTING VITAL INTELLIGENCE SIGHTINGS

TABLE OF CONTENTS

Section	Paragraph	Subject	Page
		Title Page	I
		Letter of Promulgation	III
		List of Effective Pages	V
		Record of Corrections	VII
		Table of Contents	IX - XI

CHAPTER I

GENERAL INSTRUCTIONS

	101	Purpose	1-1
	102	Scope	1-1
	103	Message Identification	1-1

CHAPTER 2

CIRVIS REPORTS

I		GENERAL	2-1
	201	Information to be Reported and When to Report	2-1
II		PROCEDURES	2-2
	202	General	2-2
	203	Precedence (priority or transmission)	2-2
	204	Content of CIRVIS Reports	2-2
	205	Additional CIRVIS Reports	2-4
	206	Addressing	2-5
	207	Acceptance of and Responsibility for CIRVIS Reports	2-7
III		SECURITY	
	208	Military and Civilian	2-8
IV		EVALUATION REPORTS	
	209	Action by Activities	2-8

IX   ORIGINAL

JANAP 146(D)

TABLE OF CONTENTS (Cont'd)

Section	Paragraph	Subject	Page
V		**SPECIAL CONSIDERATIONS**	2-9
	210	Radio Transmission Restrictions	2-9
VI		**COMMERCIAL CHARGES**	2-9
	211	Charges	2-9

CHAPTER III

MERINT REPORTS

Section	Paragraph	Subject	Page
I		**GENERAL**	3-1
	301	Information to be Reported and When to Report	3-1
II		**PROCEDURES**	3-1
	302	General	3-1
	302	Precedence (priority of transmission)	3-2
	304	Contents of MERINT Reports	3-2
	305	Additional MERINT Reports	3-3
	306	Addressing	3-4
	307	Acceptance of and Responsibility for MERINT Reports	3-6
III		**SECURITY**	3-7
	308	Military and Civilian	3-7
IV		**EVALUATION REPORTS**	3-7
	309	Action by Activities	3-7
V		**SPECIAL CONSIDERATIONS**	3-7
	310	Radio Transmission Restrictions	3-7
VI		**COMMERCIAL CHARGES**	
	311	Charges	3-7

X   ORIGINAL

JANAP 146(D)

CHAPTER I

GENERAL DESCRIPTION AND PURPOSE OF COMMUNICATION INSTRUCTIONS FOR REPORTING VITAL INTELLIGENCE SIGHTINGS

101. Purpose. - The purpose of this publication is to provide uniform instructions for the peacetime reporting of vital intelligence sightings and to provide communication instructions for the passing of these intelligence reports to appropriate military authorities.

102. Scope. -

    a. This publication is limited to the reporting of information of vital importance to the security of the United States of America and Canada and their forces, which in the opinion of the observer, requires very urgent defensive and/or investigative action by the US and/or Canadian Armed Forces.

    b. The procedures contained in this publication are provided for:

        (1) US and Canadian civil and commercial aircraft.

        (2) US and Canadian government and military aircraft other than those operating under separate reporting directives.

        (3) US and Canadian merchant vessels operating under US and Canadian registry.

        (4) US and Canadian government and military vessels other than those operating under separate reporting directives.

        (5) Certain other US and Canadian vessels including fishing vessels.

        (6) Military installations receiving reports from civilian or military land based or waterborne observers unless operating under separate reporting directives.

        (7) Government and civilian agencies which may initiate reports on receipt of information from land-based, airborne or waterborne observers.

103. Message Identification. -

    a. Reports made from airborne and land-based sources will be identified by CIRVIS pronounced SUR VEES as the first word of the text. (Refer Chapter II).

    b. Reports made by waterborne sources will be identified by MERINT pronounced as MUR ENT as the first word of the text. (Refer Chapter III).

1-1   ORIGINAL
(Reverse Blank)

JANAP 146(D)

CHAPTER II

CIRVIS REPORTS

SECTION I - GENERAL

201. <u>Information to be Reported and When to Report.</u> -

    a. Sightings within the scope of this chapter, as outlined in Article 102b(1), (2), (6) and (7), are to be reported as follows:

        (1) While airborne (except over foreign territory - see paragraph 210) and from land based observers. NOTE: Canada and the United States are not considered foreign territory for either country for the purposes of this publication.

            (a) Hostile or unidentified single aircraft or formations of aircraft which appear to be directed against the United States or Canada or their forces.

            (b) Missiles.

            (c) Unidentified flying objects.

            (d) Hostile or unidentified submarines.

            (e) Hostile or unidentified group or groups of military surface vessels.

            (f) Individual surface vessels, submarines, or aircraft of unconventional design, or engaged in suspicious activity or observed in a location or on a course which may be interpreted as constituting a threat to the United States, Canada or their forces.

            (g) Any unexplained or unusual activity which may indicate a possible attack against or through Canada or the United States, including the presence of any unidentified or other suspicious ground parties in the Polar region or other remote or sparsely populated areas.

        (2) Upon landing.

            (a) Reports which for any reason could not be transmitted while airborne.

            (b) Unlisted airfields or facilities, weather stations, or air navigation aids.

            (c) Post-landing reports.

2-1   ORIGINAL

JANAP 146(D)

SECTION II - PROCEDURES

202. <u>General</u>. - Communications procedures to be employed will be basically those prescribed for the communications system or service used. Continuing efforts will be made by an aircraft originating a CIRVIS report to insure that each CIRVIS message is received by an appropriate station.

203. <u>Precedence (priority or transmission)</u>. -

    a. To avoid delays by aircraft in rendering a CIRVIS report to a ground facility, the word "CIRVIS" spoken three (3) times will be employed, preceding the call, to clear the frequency(ies) over all other communications, except DISTRESS, URGENCY and SAFETY, to insure its expeditious handling.

    b. Should instances occur, when use of the above procedure fails to clear the frequency(ies) over all other communications in progress except as provided for in 203a, the International Urgency Signal "XXX" transmitted three (3) times or "PAN" spoken three (3) times will be employed to facilitate disposition of the message to the receiving facility.

    c. The following precedence will be employed in the transmission of all CIRVIS reports, as appropriate, commensurate with the communications facilities used:

    Tabulation

Circuit clearance	CIRVIS CIRVIS CIRVIS
International Urgency Signal (alternate)	XXX XXX XXX or PAN PAN PAN
Military precedence	Y or Emergency
Commercial class of service Indicator	RAPID US GOVT for US Government activities or RUSH for Canadian Government activities (to be used only when refiled with commercial companies)

204. <u>Contents of CIRVIS Reports</u>. -

    a. Airborne CIRVIS reports will be similar to routine aircraft position reports transmitted by either radiotelephone or radiotelegraph. The appropriate procedures to be employed will be those applicable to communications facilities utilized. The reports should contain the following information, when appropriate, in the order listed:

        (1) CIRVIS Report.

        (2) Identification of reporting aircraft or observer as appropriate.

        (3) Object sighted. Give brief description of the sighting which should contain the following items as appropriate.

JANAP 146(D)

      (a) Number of aircraft, vessels, missiles, submarines, etc.

      (b) Category of object, general description, e.g., size, shape, type of propulsion, etc.

  (4) The position of the object. This can be indicated by any of the following methods:

      (a) Latitude and Longitude.

      (b) Over a radio fix.

      (c) True bearing and distance from a radio fix.

      (d) Over a well-known or well-defined geographic point.

      (e) True bearing and distance from a geographic point.

  (5) Date and time of sighting (GMT).

  (6) Altitude of object.

  (7) Direction of travel of object.

  (8) Speed of object.

  (9) Any observed identification, insignia, or other significant information. Every reasonable effort should be made to positively identify the object sighted.

Example of an air/ground radiotelephone transmission:

(Aircraft)   CIRVIS CIRVIS CIRVIS - KINDLEY THIS IS AIR FORCE TWO FIVE NINE THREE SIX - CIRVIS REPORT - OVER

(Aeronautical Station)   AIR FORCE TWO FIVE NINE THREE SIX THIS IS KINDLEY - GO AHEAD

(Aircraft)   EMERGENCY - CIRVIS REPORT - AIR FORCE TWO FIVE NINE THREE SIX SIGHTED FORMATION OF SIX JET BOMBERS - CONFIGURATION IS SWEPT WING WITH EIGHT JET ENGINES - TWO HUNDRED MILES EAST OF BERMUDA ON THIRTEEN MAY AT ONE THREE FIVE ZERO ZULU - ALTITUDE THREE FIVE THOUSAND - HEADING TWO SEVEN ZERO DEGREES - NO MARKINGS OBSERVED - OVER

(Aeronautical Station)   KINDLEY - ROGER - OUT

2-3   ORIGINAL

JANAP 146(D)

Example of an air/ground radiotelegraph transmission:

(Aircraft)   XXX XXX XXX AFA3 DE A48207

(Aeronautical
  Station)   A48207 DE AFA3 K

(Aircraft)   Y - CIRVIS REPORT. A48207 SIGHTED ....ETC.

(Aeronautical
  Station)   A48207 DE AFAR AR

205. Additional CIRVIS Reports. -

    a. Additional reports should be made if more information becomes available concerning a previously sighted object. These reports should contain a reference to the original report sufficient to identify them with the original sighting.

Example of an air/ground radiotelephone transmission:

(Aircraft)   CIRVIS CIRVIS CIRVIS - KINDLEY THIS IS AIR FORCE TWO
                          FIVE NINE THREE SIX - CIRVIS REPORT - OVER

(Aeronautical
  Station)   AIR FORCE TWO FIVE NINE THREE SIX - THIS IS KINDLEY -
                          GO AHEAD

(Aircraft)   EMERGENCY - THE SIX JET BOMBERS PREVIOUSLY REPORTED AT
                          ONE THREE FIVE ZERO ZULU BY AIR FORCE TWO FIVE NINE
                          THREE SIX ARE NOW ONE THREE ZERO MILES WEST OF BERMUDA
                          AT ONE FOUR THREE FIVE ZULU - HEADING TWO SEVEN ZERO
                          DEGREES - OVER

(Aeronautical
  Station)   KINDLEY - ROGER - OUT

NOTE: In radiotelegraph transmission, the same procedures would apply as prescribed in para 204.

    b. Cancellation reports should be made in the event a previously reported sighting is positively identified as friendly or that it has been erroneously reported. Such reports should be transmitted as a brief message cancelling the previous report(s).

Example of an air/ground radiotelephone transmission:

(Aircraft)   CIRVIS CIRVIS CIRVIS - KINDLEY THIS IS AIR FORCE TWO
                          FIVE NINE THREE SIX - CIRVIS REPORT - OVER

(Aeronautical
  Station)   AIR FORCE TWO FIVE NINE THREE SIX THIS IS KINDLEY -
                          GO AHEAD

2-4   ORIGINAL

JANAP 146(D)

(Aircraft)   EMERGENCY - CANCEL CIRVIS REPORT OF ONE THREE FIVE ZERO
                   ZULU BY AIR FORCE TWO FIVE NINE THREE SIX - SIX JET
                   BOMBERS POSITIVELY IDENTIFIED AS AIR FORCE BRAVO FORTY
                   SEVENS AT ONE FOUR FOUR SIX ZULU - OVER

(Aeronautical
  Station)   KINDLEY - ROGER - OUT

NOTE: In radiotelegraph transmission, the same procedures would apply
      as prescribed in para 204.

    c. A post-landing report is desired immediately after landing
by CINCNORAD or RCAF-ADC to amplify the airborne report(s). This may
be filed with either the military or civil communications facility
located at the place of landing. If the landing is not made in Canadian or United States territory the report should be made to the nearest
Canadian or United States military or diplomatic representative in that
area. The post-landing report will refer to the airborne report(s) and,
in addition, contain a brief resume of weather conditions at the time
of sighting(s), verification of the sighting(s) by other personnel and
any other information deemed appropriate. If the sighting was identified
as friendly, and a report so stating was filed while airborne, no post-
landing report is required.

        (1) If no airborne report was made as a result of inability
            to reach a communications station or due to being over
            foreign territory (see paragraph 210), the post-landing
            report will contain all the information available concerning the sighting.

206. <u>Addressing.</u> -

    a. Aircraft. - It is imperative that all CIRVIS reports reach
the appropriate military commands as quickly as possible. The reports,
therefore, shall be transmitted as soon as possible after the sighting.
Ground procedures have been established to handle CIRVIS reports by
either military or civil facilities, so the same procedures as those
now established and in use by pilots for air traffic control shall be
followed. When contact by civil or military pilots cannot be established with any ground communications station, maximum effort shall be
made to relay the CIRVIS reports via other aircraft with which communication is possible.

        (1) Post-landing reports should be addressed to CINCNORAD,
            Ent AFB, Colorado Springs, Colorado, or RCAF-ADC, St.
            Hubert, Quebec whichever is the more convenient <u>if the
            sighting occurred within or adjacent to the North
            American continent</u>. Whichever of these headquarters
            receives the report will immediately notify the other
            and also all other addressees of the original report(s).
            If the sighting(s) occurred in other areas, the post-
            landing report should be made to the nearest US or

2-5   ORIGINAL

JANAP 146(D)

Canadian military or diplomatic representative in that area who will forward the report as prescribed in subparagraph 206b(1)(a).

 b. Communications Stations. - Communications stations (to include any civil or military facility such as control tower, naval shore radio station, approach control, ARTC center, or any other communications facility) receiving CIRVIS reports will immediately after receipting process the report as follows (for <u>additional</u> instructions to US military fixed communications stations in Canada, Alaska and Greenland see subparagraph (2) (a) below):

 (1) US military fixed communications stations will multiple-address the CIRVIS report to the following address designations:

  (a) For sightings in overseas areas - reports will be forwarded to:

   <u>1</u>. Addressees as prescribed by Area Commanders. (Normally, these addressees are the operating service commands concerned).

   <u>2</u>. Commander-in-Chief, North American Air Defense Command (CINNORAD), Ent AFB, Colorado Springs, Colorado.

   <u>3</u>. Chief of Staff, United States Air Force (COFS, USAF), Washington, D. C.

 (2) Canadian and US military fixed communications stations will multiple address the CIRVIS reports to the following address designations:

  (a) For sightings within or adjacent to the North American continent, reports will be forwarded to:

   <u>1</u>. Commander of the nearest joint air defense division, command or group.

   <u>2</u>. CINCNORAD, Ent AFB, Colorado Springs, Colorado.

   <u>3</u>. Appropriate Sea Frontier Command:

    <u>a</u>. Commander, Western Sea Frontier (COMWESTSEAFRON), San Francisco, California.

    <u>b</u>. Commander, Eastern Sea Frontier (COMEASTSEAFRON), New York, N. Y.

   <u>4</u>. Chief of Staff, United States Air Force (COFS, USAF) Washington, D. C.

2-6   ORIGINAL

JANAP 146(D')

        5. RCAF Air Defense Command (CANAIRDEF) St. Hubert, Montreal, Canada.

        6. Appropriate Flag Officer in Command:

           a. Canadian Flag Officer, Atlantic Coast, (CANFLAGLANT), Halifax, Nova Scotia.

           b. Canadian Flag Officer, Pacific Coast, (CANFLAGPAC), Esquimalt, British Columbia.

    (3) Civil communications stations will handle CIRVIS reports received from either aircraft or other communications stations as follows:

        (a) Air Carrier company stations will pass the CIRVIS report, exactly as received, to the nearest CAA or DOT ARTC center in the same manner as air traffic control information.

        (b) CAA or DOT communications stations, upon receipt of a CIRVIS report will immediately pass the report to the appropriate ARTC center.

        (c) CAA or DOT ARTC Centers. Upon receipt of CIRVIS reports, ARTC centers will forward them immediately to the appropriate military facility as prescribed by agreement with the appropriate military commander.

207. Acceptance of and Responsibility for CIRVIS Reports. -

    a. The following activities have responsibilities as follows:

       (1) CONCNORAD or RCAF-ADC will review all CIRVIS reports to ascertain that they have been addressed in accordance with paragraph 206 and forward reports to any omitted addressees in the United States and Canada respectively. These headquarters are the normal points of contact between the two countries and are responsible for passing CIRVIS reports of interest, including post-landing reports, to each other.

       (2) United States or Canadian military or diplomatic authorities in receipt of CIRVIS reports that have not been previously forwarded should take the action indicated in paragraph 206 without delay by the most rapid means available.

       (3) Chief of Staff, USAF, will disseminate CIRVIS reports to appropriate agencies in the Washington, D. C. area.

       (4) RCAF-ADC and the Canadian Flag Officers will be responsible for notifying Canadian military headquarters in Ottawa concerning CIRVIS reports.

JANAP 146(D)

(5) Sea Frontier Commanders will be responsible for notifying Chief of Naval Operations and appropriate Fleet Commanders concerning CIRVIS reports.

    b. Fixed and mobile military communications facilities and military personnel having occasion to handle CIRVIS reports must lend assistance in all cases required in expediting CIRVIS reports. All civilian facilities and personnel are also urged to do so. Maximum effort must be made by all persons handling CIRVIS reports to insure positive immediate delivery.

    c. WHEN A STATION RECEIVES A PARTIAL CIRVIS REPORT AND THE REMAINDER IS NOT IMMEDIATELY FORTHCOMING, IT WILL BE RELAYED OR DELIVERED IN THE SAME MANNER AS A COMPLETE REPORT.

SECTION III - SECURITY

208. **Military and Civilian.** - Transmission of CIRVIS reports are subject to the U. S. Communications Act of 1934, as amended, and the Canadian Radio Act of 1938, as amended. Any person who violates the provisions of these acts may be liable to prosecution thereunder. These reports contain information affecting the National Defense of the United States and Canada. Any person who makes an unauthorized transmission or disclosure of such a report may be liable to prosecution under Title 18 of the US Code, Chapter 37, or the Canadian Official Secrets Act of 1939, as amended. This should not be construed as requiring classification of CIRVIS messages. The purpose is to emphasize the necessity for the handling of such information within official channels only.

SECTION IV - EVALUATION REPORTS

209. **Action by Activities.** -

    a. All investigative measures and evaluation processes instituted by addressees, and by originating authorities where applicable, will be handled in accordance with existing procedures and reported in accordance with these instructions, insuring that appropriate commands as listed in paragraph 206 are kept fully informed of investigative results and evaluations. These evaluations shall be expressed in terms indicating the reported sighting as being Positive, Probable, Possible, or No Threat <u>insofar as being a threat to the security of the United States of America or Canada or their forces</u>, or an explanation of the subject reported when known.

    b. The first two words of the text of an evaluation report shall be "CIRVIS EVALUATION" followed by the date-time-group and/or other identification of the CIRVIS report(s) being evaluated.

JANAP 146(D)

SECTION V - SPECIAL CONSIDERATIONS

210. <u>Radio Transmission Restrictions</u>. - CIRVIS reports will not be transmitted by radio while over foreign territory, other than Greenland or Iceland, but will be transmitted as soon as practicable upon leaving foreign territorial boundaries. In accordance with special permission from the Danish government, reports may be transmitted while traversing Greenland. Foreign territory includes all territory except international water areas and territory under the jurisdiction of the United States of America and Canada.

SECTION VI - COMMERCIAL CHARGES

211. <u>Charges</u>. -

    a. All charges incurred in handling CIRVIS reports through U. S. facilities will be charged to the U. S. Department of the Air Force (accounting symbol "AF"). Insofar as practicable, CIRVIS reports so handled should be forwarded <u>RAPID US GOVT COLLECT</u>.

    b. All charges incurred in handling CIRVIS reports through Canadian facilities will be charged to the Royal Canadian Air Force. Insofar as practicable, CIRVIS reports so handled will be forwarded "RUSH COLLECT".

    c. Any or all questions of charges will be resolved after traffic has been handled. In no case should CIRVIS reports be delayed because of communication handling charges.

2-9   ORIGINAL
(Reverse Blank)

JANAP 146(D)

CHAPTER III

MERINT REPORTS

SECTION 1 - GENERAL

301. <u>Information to be Reported and When to Report</u>. -

    a. Sightings within the scope of this chapter, as outlined in Article 102b, (3), (4), (5) and (6) are to be reported as follows:

        (1) Immediately (except when within territorial waters of nations other than Canada or the USA as prescribed by International Law).

            (a) Hostile or unidentified single aircraft or formation of aircraft which appear to be directed against Canada or the United States or their forces.

            (b) Missiles.

            (c) Unidentified flying objects.

            (d) Hostile or unidentified submarines.

            (e) Hostile or unidentified group or groups of military surface vessels.

            (f) Individual surface vessels, submarines, or aircraft of unconventional design, or engaged in suspicious activities or observed in an unusual location.

            (g) Any unexplained or unusual activity which may indicate possible attack against or through Canada or the United States, including the presence of any unidentified or other suspicious ground parties in the Polar Region or other remote or sparsely populated areas.

SECTION II - PROCEDURES

302. <u>General</u>. - Communication procedures to be employed will be basically those prescribed for the communications system or services used. Merchant ships will employ normal international commercial communication procedures and utilize existing commercial or military facilities as appropriate. Every effort will be made to obtain an acknowledgment for each MERINT message transmitted. Canadian or U. S. vessels which are manned by military or civil service personnel will use military communication procedure.

3-1   ORIGINAL

JANAP 146(D)

303. <u>Precedence (priority of transmission)</u>. - Transmission of MERINT reports shall be preceded by the word "MERINT" spoken three times OR by its alternate, the international "Urgency Signal". Additionally, the military precedence of "Emergency" shall be used if the report is addressed to military activities. Governmental precedence of "Rapid U. S. Government" for reports addressed to other U. S. Government activities, or Canadian "Rush", for reports addressed to Canadian Government activities shall be used:

<u>Tabulation</u>

Circuit clearance	MERINT MERINT MERINT
International Urgency Signal (Alternate)	XXX XXX XXX or PAN PAN PAN
Military Precedence	Y or Emergency
Commercial Class of Service Indicator	RAPID US GOVT for US Government activities or RUSH for Canadian Government activities (to be used only when refiled with commercial companies)

304. <u>Contents of MERINT Reports</u>. -

    a. MERINT reports should contain the following as applicable in the order listed:

    (1) "MERINT" will always be the first word of the text.

    (2) Name and call letters of reporting ship.

    (3) Object sighted. Give brief description of the sighting which should contain the following items as appropriate:

        (a) Number of aircraft, vessels, missiles, submarines, etc.

        (b) Category of object, general description, e.g., size, shape, type of propulsion, etc.

    (4) Ship's position at time of sightings.

    (5) Date and time of sighting (GMT)*

    (6) Altitude of object expressed as Low, Medium or High.

    (7) Direction of travel of object.

    (8) Speed of object.

    (9) Any observed identification, insignia, or other significant information. Every reasonable effort should be made to positively identify the object sighted.

JANAP 146(D)

(10) Conditions of sea and weather.

* "071430Z" is an example of a complete date-time group (DTG). When broken into component parts (07) is the day of the month, followed by (14) the hour in 24 hour time, followed by (30) the minutes of the hour, followed by (Z) the time zone. "Z" signifies that Greenwich Mean Time has been used in composing the date-time group.

| Day of Month | Hour Expressed in 24 hour time | Minutes of the hour | Indication that GMT is being used |

EXAMPLE of a Radiotelephone Transmission:

MERINT MERINT MERINT - WHISKEY ZULU TANGO - THIS IS KILO HOTEL WHISKEY MIKE - OVER
KILO HOTEL WHISKEY MIKE - THIS IS - WHISKEY ZULU TANGO - OVER
WHISKEY ZULU TANGO - THIS IS - KILO HOTEL WHISKEY MIKE
MERINT SS TUNA KILO HOTEL WHISKEY MIKE SIGHTED FORMATION OF SIX JET BOMBERS LAT 40N 50E AT 211500Z ALTITUDE MEDIUM HEADING 270 DEGREES TYPE OF AIRCRAFT NOT OBSERVED BEFORE WIND FORCE 3 SEA CALM -
OVER

EXAMPLE of a Radiotelegraph Transmission:

MERINT MERINT MERINT CFH DE KHWM K
KHWM DE CFH K
CFH DE KHWM
"RAPID U S GOVERNMENT" or CANADIAN "RUSH"
MERINT (REMAINDER OF TEXT)
211513Z JONES KHWM
K

305. Additional MERINT Reports. -

    a. Amplifying Reports. -

        (1) When additional information becomes available to any observer and is of importance, it is to be transmitted as a "MERINT AMPLIFY" report.

        (2) Amplifying reports are to be handled in the same way as the original report except that the first two words of the text will be "MERINT AMPLIFY" followed by the date and time of filing of the MERINT report being amplified.

        (3) Amplifying reports on aerial objects normally consist of additional information pertaining to the sighted object's size, shape; description of jet or rocket streams; color, sound; if multiple objects, the number; type; method of propulsion; number of engines; insignia and estimated speed.

3-3   ORIGINAL

JANAP 146(D)

(4) Amplifying reports on submarines or surface sightings normally consist of additional information on the state of sea and weather, including visibility; actions of object (course, speed, evasive maneuvers, etc.) identification marks, (flags, signals, numbers, exchange of communication); closest range at which object observed; any further remarks, (dived, surfaced, commenced snorkling, stopped snorkling, etc.)

b. Cancellation Reports. -

(1) Cancellation reports should be made in the event a previously reported sighting is positively identified as friendly, erroneous or false.

(2) MERINT cancellations are to be handled in the same manner as the original MERINT report except that the first two words shall be "MERINT CANCEL" followed by the date and time of filing of the MERINT report being cancelled and, in brief, the information on which the cancellation is based.

c. Delayed Reports. - In the event a MERINT report cannot be made by radio, the Master is requested to report the details of the MERINT sightings to the appropriate Canadian or United States military authorities. If the port of arrival is outside of Canada or USA, the report is to be made to the nearest Canadian or US military or diplomatic representative in the area. This report should be submitted immediately upon arrival in port by any available means and should be in the format prescribed in paragraph 304.

306. Addressing. -

a. Vessels. -

(1) It is imperative that all MERINT reports reach the appropriate military commands as quickly as possible. The reports, therefore, shall be transmitted as soon as possible after the sighting. All Canadian or U.S. vessels listed under Para 102b, (3), (4), and (5) are to transmit in plain language all MERINT reports to the nearest Canadian or U. S. military, government or commercial radio station, regardless of whether the vessel is Canadian or U. S. registered.

(2) Naval vessels which intercept MERINT reports from merchant vessels shall copy the report and immediately relay EXACTLY AS RECEIVED to the appropriate Canadian or U. S. Navy-Shore Radio Station with relay instructions.

3-4 ORIGINAL

JANAP 146(D)

b. Communications Stations. - Communications Stations (to include any commercial, government or military facility such as coastal marine radio station, telegraph offices, naval or coast guard shore radio station or any other communication facility) receiving MERINT reports will immediately after receipting process the report as follows:

(1) Canadian or U. S. commercial or government communications stations will handle MERINT reports received from either vessels or other communications stations by immediately forwarding them to a Canadian or U. S. military fixed communication facility as prescribed by agreement with the appropriate military commander.

(2) U. S. military fixed communications stations will multiple-address the MERINT report to the following address designations:

(a) For sightings in overseas areas - reports will be forwarded to:

1. Addressees as prescribed by Area Commanders. (Normally, these addressees are the operating Service commands concerned).

2. Commander-in-Chief. North American Air Defense Command (CINCNORAD), Ent AFB, Colorado Springs,

3. Chief of Staff, United States Air Force (COFS, USAF), Washington, D. C.

(3) Canadian and U. S. military communications stations will multiple-address the MERINT reports to the following address designations:

(a) For sightings within or adjacent to the North American continent, reports will be forwarded to:

1. Commander of the nearest joint air defense division, command or group.

2. CINCNORAD, Ent AFB, Colorado Springs, Colorado.

3. Appropriate Sea Frontier Command:

a. Commander, Western Sea Frontier (COMWESTSEAFRON), San Francisco, Calif.

b. Commander, Eastern Sea Frontier (COMEASTSEAFRON), New York, N. Y.

4. Chief of Staff, United States Air Force, (COFS USAF), Washington, D. C.

3-5 ORIGINAL

JANAP 146(D)

  <u>5</u>. RCAF Air Defense Command (CANAIRDEF), St. Hubert, Montreal.

  <u>6</u>. Appropriate Flag Officer in Command:

    <u>a</u>. Canadian Flag Officer, Atlantic Coast, (CANFLAGLANT), Halifax, Nova Scotia.

    <u>b</u>. Canadian Flag Officer, Pacific Coast, (CANFLAGPAC), Esquimalt, British Columbia.

307. <u>Acceptance of and Responsibility for MERINT Reports</u>. -

  a. The following activities have responsibilities as follows:

   (1) CINCNORAD or RCAF-ADC will review all MERINT reports to ascertain that such reports have been addressed in accordance with paragraph 306 and forward reports to any omitted addressees in U. S. and Canada respectively. These headquarters are the normal points of contact between the two countries and are responsible for passing MERINT reports of interest, including delayed reports, to each other.

   (2) Canadian or United States military or diplomatic authorities in receipt of MERINT reports will take the action indicated in paragraph 306 without delay by the most rapid means available.

   (3) Chief of Staff, USAF, will disseminate MERINT reports to appropriate agencies in the Washington, D. C. area.

   (4) RCAF-ADC and the Canadian Flag Officers will be responsible for notifying Canadian military headquarters in Ottawa concerning MERINT reports.

   (5) Sea Frontier Commanders will be responsible for notifying Chief of Naval Operations and the appropriate Fleet Commanders concerning MERINT reports.

  b. Fixed and mobile military communications facilities and military personnel having occasion to handle MERINT reports must lend assistance in all cases required in expediting MERINT reports. All civilian facilities and personnel are also urged to do so. Maximum effort should be made by all persons handling MERINT reports to insure positive immediate delivery.

  c. WHEN A STATION RECEIVES A PARTIAL MERINT REPORT AND THE REMAINDER IS NOT IMMEDIATELY FORTHCOMING, IT WILL BE RELAYED OR DELIVERED IN THE SAME MANNER AS A COMPLETE REPORT.

JANAP 146(D)

SECTION III - SECURITY

308. <u>Military and Civilian</u>. - Transmission of MERINT reports are subject to the Communications Act of 1934, as amended, and the Canadian Radio Act of 1938, as amended. Any person who violates the provisions of these acts may be liable to prosecution thereunder. These reports contain information affecting the National Defense of the United States and Canada. Any person who makes an unauthorized transmission or disclosure of such a report may be liable to prosecution under Title 18 of the US Code, Chapter 37, or the Canadian Official Secrets Act of 1939, as amended. This should not be construed as requiring classification of MERINT messages. The purpose is to emphasize the necessity for the handling of such information within official channels only.

SECTION IV - EVALUATION REPORTS

309. <u>Action by Activities</u>. -

    a. All investigative measures and evaluation processes instituted by addressees, and by originating authorities, where applicable, will be handled and reported in accordance with existing procedures, insuring that appropriate commands as listed in paragraph 306 are kept fully informed of investigative results and evaluations. These evaluations shall be expressed in terms indicating the reported sighting as being Positive, Probable, Possible or No Threat <u>insofar as being a threat to the security of the United States of America or Canada or their forces</u>, or an explanation of the subject reported when known.

    b. The first two words of the text of an evaluation report shall be "MERINT EVALUATION" followed by the date-time-group and/or other identification of the MERINT report(s) being evaluated.

SECTION V - SPECIAL CONSIDERATIONS

310. <u>Radio Transmission Restrictions</u>. - MERINT reports will not be transmitted by radio other than Canadian or U. S. or international waters.

SECTION VI - COMMERCIAL CHARGES

311. <u>Charges</u>. -

    a. All charges incurred in handling MERINT reports through U. S. facilities will be charged to the U. S. Department of the Navy (accounting symbol NAVY). Insofar as practicable, MERINT reports so handled should be forwarded <u>RAPID US GOVT COLLECT</u>.

    b. All charges incurred in handling MERINT reports through facilities will be charged to the Royal Canadian Navy. Insofar

3-7   ORIGINAL

JANAP 146(D)

as practicable, MERINT reports so handled will be forwarded "RUSH COLLECT".

c. Any or all questions of charges will be resolved after traffic has been handled. In no case should MERINT reports be delayed because of communication handling charges.

ORIGINAL

NEWS RELEASE
PLEASE NOTE DATE

DEPARTMENT OF DEFENSE
OFFICE OF PUBLIC INFORMATION
Washington 25, D. C.

IMMEDIATE RELEASE   OCTOBER 25, 1955   NO. 1053-55
LI 5-6700, Ext 75131

AIR FORCE RELEASES STUDY ON UNIDENTIFIED AERIAL OBJECTS

The results of an investigation begun by the Air Force in 1947 into the field of Unidentified Aerial Objects (so-called flying saucers) were released by the Air Force today.

No evidence of the existence of the popularly-termed "flying saucers' was found.

The report was based on study and analysis by a private scientific group under the supervision of the Air Technical Intelligence Center at Dayton, Ohio. Since the instigation of the investigation more than seven years ago, methods and procedures have been so refined that of the 131 sightings reported during the first four months of 1955 only three per cent were listed as unknown. (A summary of the report is attached.)

Commenting on this report, Secretary of the Air Force Donald A. Quarles said: "On the basis of this study we believe that no objects such as those popularly described as flying saucers have overflown the United States. I feel certain that even the unknown three per cent could have been explained as conventional phenomena or illusions if more complete observational data had been available.

"However, we are now entering a period of aviation technology in which aircraft of unusual configuration and flight characteristics will begin to appear.

"The Air Force and the other Armed Services have under development several vertical-rising, high performance aircraft, and as early as last year a propeller driven vertical-rising aircraft was flown. The Air Force will fly the first jet-powered vertical-rising airplane in a matter of days. We have another project under contract with AVRO Ltd., of Canada, which could result in disc-shaped aircraft somewhat similar to the popular concept of a flying saucer. An available picture, while only an artists' conception, could illustrate such an object. (Photograph is available at Pictorial Branch, Room 2D780, Ext. 75331).

"While some of these may take novel forms, such as the AVRO project, they are direct-line descendents of conventional aircraft and should not be regarded as supra-natural or mysterious. We expect to develop airplanes that will fly faster, higher and perhaps farther than present designs, but they will still obey natural laws and if manned, they will still be manned by normal terrestrial airmen. Other than reducing runway requirements we do not expect vertical-rising aircraft to have more outstanding military characteristics than conventional types.

MORE

"Vertical-rising aircraft capable of transition to supersonic horizontal flight will be a new phenomenon in our skies, and under certain conditions could give the illusion of the so-called flying saucer. The Department of Defense will make every effort within bounds of security to keep the public informed of these developments so they can be recognized for what they are."

Mr. Quarles added: "I think we must recognize that other countries also have the capability of developing vertical-rising aircraft, perhaps of unconventional shapes. However we are satisfied at this time that none of the sightings of so-called 'flying saucers' reported in this country were in fact aircraft of foreign origin."

E N D

Attachment

SUMMARY

(Analysis Of Reports Of Unidentified Aerial Objects)

Reports of unidentified aerial objects (popularly termed "flying saucers" or "flying discs") have been received by the U.S. Air Force since mid-1947 from many and diverse sources. Although there was no evidence that the unexplained reports of unidentified objects constituted a threat to the security of the United States, the Air Force determined that all reports of unidentified aerial objects should be investigated and evaluated to determine if "flying saucers" represented technological developments not known to this country.

In order to discover any pertinent trend or pattern inherent in the data, and to evaluate or explain any trend or pattern found, appropriate methods of reducing these data from reports of unidentified aerial objects to a form amenable to scientific appraisal were employed. In general, the original data upon which this study was bases consisted of impressions and interpretations of apparently unexplainable events, and seldom contained reliable measurements of physical attributes. This subjectivity of the data presented a major limitation to the drawing of significant conclusions, but did not invalidate the application of scientific methods of study.

The reports received by the U.S. Air Force on unidentified aerial objects were reduced to IBM punched-card abstracts of data by means of logically developed forms and standardized evaluation procedures. Evaluation of sighting reports, a crucial step in the preparation of the data for statistical treatment, consisted of an appraisal of the reports and the subsequent categorization of the object or objects described in each report. A detailed description of this phase of the study stresses the careful attempt to maintain complete objectivity and consistency.

Analysis of the refined and evaluated data derived from the original reports of sightings consisted of (1) a systematic attempt to ferret out any distinguishing characteristics inherent in the data of any of their segments, (2) a concentrated study of any trend or pattern found, and (3) an attempt to determine the probability that any of the UNKNOWNS represent observations of technological developments not known to this country.

The first step in the analysis of the data revealed the existence of certain apparent similarities between cases of objects definitely identified and those not identified. Statistical methods of testing when applied indicated a low probability that these apparent similarities were significant. An attempt to determine the probability that any of the UNKNOWNS represented observations of technological developments not known to this country necessitated a thorough re-examination and re-evaluation of the cases of objects not originally identified; this led to the conclusion that this probability was very small.

MORE

The special study which resulted in this report (Analysis of Reports of Unidentified Aerial Objects, 5 May 1955) started in 1953. To provide the study group with a complete set of files, the information cut-off date was established as of the end of 1952. It will accordingly be noted that the statistics contained in all charts and tables in this report are terminated with the year 1952. In these charts, 3201 cases have been used.

As the study progressed, a constant program was maintained for the purpose of making comparisons between the current cases received after 1 January 1953, and those being used for the report. This was done in order that any change or significant trend which might arise from current developments could be incorporated in the summary of this report.

The 1953 and 1954 cases show a general and expected trend of increasing percentages in the finally identified categories. They also show decreasing percentages in categories where there was insufficient information and those where the phenomena could not be explained. This trend had been anticipated in the light of improved reporting and investigating procedures.

Official reports on hand at the end of 1954 totaled 4834. Of these, 425 were produced in 1953 and 429 in 1954. These 1953 and 1954 individual reports (a total of 854), were evaluated on the same basis as were those received before the end of 1952. The results are as follows:

```
          Balloons ..................... 16 per cent
          Aircraft ..................... 20 per cent
          Astronomical ................. 25 per cent
          Other ........................ 13 per cent
          Insufficient Information ..... 17 per cent
          Unknown ......................  9 per cent
```

As the study of the current cases progressed, it became increasingly obvious that if reporting and investigating procedures could be further improved, the percentages of those cases which contained insufficient information and those remaining unexplained would be greatly reduced. The key to a higher percentage of solutions appeared to be in rapid "on the spot" investigations by trained personnel. On the basis of this, a revised program was established by Air Force Regulation 200-2, Subject: "Unidentified Flying Objects Reporting" (Short Title: UFOB), dated 12 August 1954.

This new program, which had begun to show marked results before January 1955, provided primarily that the 4602d Air Intelligence Service Squadron (Air Defense Command) would carry out all field investigations. This squadron has sufficient units and is so deployed as to be able to arrive "on the spot" within a very short time after a report is received. After treatment by the 4602d Air Intelligence Service Squadron, all information is supplied to the Air Technical Intelligence Center for final evaluation. This cooperative program has resulted, since 1 January 1955, in reducing the insufficient information cases to seven percent and the unknown cases to three percent of the totals.

MORE

The period 1 January 1955 to 5 May 1955 accounted for 131 unidentified aerial object reports received. Evaluation percentages of these are as follows:

```
            Balloons ........................ 26 per cent
            Aircraft ........................ 21 per cent
            Astronomical .................... 23 per cent
            Other ........................... 20 per cent
            Insufficient Information ........  7 per cent
            Unknown .........................  3 per cent
```

All available data were included in this study which was prepared by a panel of scientists both in and out of the Air Force. On the basis of this study it is believed that all the unidentified aerial objects could have been explained if more complete observational data had been available. Insofar as the reported aerial objects which still remain unexplained are concerned, there exists little information other than the impressions and interpretations of their observers. As these impressions and interpretations have been replaced by the use of improved methods of investigation and reporting, and by scientific analysis, the number of unexplained cases has decreased rapidly towards the vanishing point.

Therefore, on the basis of this evaluation of the information, it is considered to be highly improbable that reports of unidentified aerial objects examined in this study represent observations of technological developments outside of the range of present-day scientific knowledge. It is emphasized that there has been a complete lack of any valid evidence of physical matter in any case of a reported unidentified aerial object.

E N D

NEWS RELEASE
PLEASE NOTE DATE

DEPARTMENT OF DEFENSE
OFFICE OF PUBLIC INFORMATION
Washington 25, D. C.

FACT SHEET   NOVEMBER 5, 1957   NO. 1083-58

LI 5-6700 Ext. 75131

AIR FORCE'S 10 YEAR
STUDY OF UNIDENTIFIED FLYING OBJECTS

In response to queries as to results of previous investigation of Unidentified Flying Object reports, the Air Force said today that after 10 years of investigation and evaluation of UFO's, no evidence has been discovered to confirm the existence of so-called "Flying Saucers."

Reporting, investigation, analysis and evaluation procedures have improved considerably since the first sighting of a "flying saucer" was made on 27 June 1947. The study and analysis of reported sightings of UFO's is conducted by a selected scientific group under the supervision of the Air Force.

Dr. J. Allen Hynek, Professor of Astrophysics and Astronomy at Ohio State University, is the Chief Scientific Consultant to the Air Force on the subject of Unidentified Flying Objects.

The selected, qualified scientists, engineers, and other personnel involved in these analyses are completely objective and open minded on the subject of "flying saucers." They apply scientific methods of examination to all cases in reaching their conclusions. The attempted identification of the phenomenon observed is generally derived from human impressions and interpretations and not from scientific devices or measurements. The data in the sightings reported are almost invariably subjective in nature. However, no report is considered unsuitable for study and categorization and no lack of valid evidence of physical matter in the "case studies is assumed to be "prima facie" evidence that so-called "flying saucers" or interplanetary vehicles do not exist.

General categories of identification are balloons, aircraft, astronomical, other, insufficient data and unknowns.

Approximately 4,000 balloons are released in the U. S. every day. There are two general types of balloons: weather balloons and upper-air research balloons. Balloons will vary from small types 4 feet in diameter to large types 200 feet in diameter. The majority released at night carry running lights which often contribute to weird or unusual appearances when observed at night. This also holds true when observed near dawn or sunset because of the effect of the slant rays of the sun upon the balloon surfaces. The large balloons, if caught in jet streams, may assume a near horizontal position when partially inflated, and move with speeds of over 200 MPH. Large types may be

MORE

observed flattened on top. The effect of the latter two conditions can be startling even to experienced pilots.

Many modern aircraft, particularly swept and delta wing types, under adverse weather and sighting conditions are reported as unusual objects and "flying saucers." When observed at high altitudes, reflecting sunlight off their surfaces, or when only their jet exhausts are visible at night, aircraft can have appearances ranging from disc to rocket in shape. Single jet bombers having multi-jet pods under their swept-back wings have been reported as UFOs or "saucers" in "V" formation. Vapor trails will often appear to glow with fiery red or orange streaks when reflecting sunlight. Afterburners are frequently reported as UFOs.

The astronomical category includes bright stars, planets, meteors, comets, and other celestial bodies. When observed through haze, light fog, or moving clouds, the planets Venus, Mars, and Jupiter have often been reported as unconventional, moving objects. Attempts to observe astronomical bodies through hand-held binoculars under adverse sky conditions has been a source of many UFO reports.

The "other" category includes reflections, searchlights, birds, kites, blimps, clouds, sun-dogs, spurious radar indications, hoaxes, firework displays, flares, fireballs, ice crystals, bolides, etc., as examples: Large Canadian geese flying low over a city at night, with street lights reflecting off their bodies; searchlights playing on scattered clouds, appearing as moving disc-like shapes.

The insufficient data category include all sightings where essential or pertinent items of information are missing, making it impossible to form a valid conclusion. These include description of the size, shape or color of the object; direction and altitude; exact time and location; wind weather conditions, etc. This category is not used as a convenient way to get rid of what might be referred to as "unknowns". However, if the data received is insufficient or unrelated, the analysts must then place that particular report in this category. The Air Force needs complete information to reach a valid conclusion. Air Force officials stressed the fact that an observer should send a complete report of a bona fide sighting to the nearest Air Force activity. There the report will be promptly forwarded to the proper office for analysis and evaluation.

A sighting is considered an "unknown" when a report contains all pertinent data necessary to normally suggest at least one valid hypothesis on the cause or explanation of the sighting but when the description of the object and its maneuvers cannot be correlated with any known object or phenomenon. In its Project Blue Book Special Report #14, released in October, 1955, the Air Force showed that evaluated sightings in the "unknown" category had been reduced to 3 percent at that time.

Previously "unknown" sightings had been 9% in 1953 and 1954 and in the previous years "flying saucer" sightings had run as high as 20% "unknowns." Project Blue Book Special Report #14, covered "flying saucer" investigations from June 1947 to May 1955. Latest Air Force statistics show the number of unknowns has since been reduced to less than 2%.

The following table presents the results of the evaluation of all reports received by the Air Force from the time that Project Blue Book, Special Report #14, was completed through June 1957. The table gives the percentage of all the reports received by the Air Force during each time period.

	1955 June thru December	1956	1957 January thru June
Balloons	27.4%	26.0%	26.4%
Aircraft	29.3%	24.6%	28.8%
Astronomical	20.1%	26.3%	24.4%
Other (Hoax, searchlight, birds, etc)	12.3%	6.8%	6.4%
Insufficient Information	8.8%	14.1%	12.1%
Unknown	2.1%	2.2%	1.9%
TOTAL NUMBER OF SIGHTINGS	273	778	250

Air Force conclusions for the ten years of UFO sightings involving approximately 5,700 reports were: first, there is no evidence that the "unknowns" were inimical or hostile; second, there is no evidence that these "unknowns" were interplanetary space ships; third, there is no evidence that these unknowns represented technological developments or principles outside the range of our present day scientific knowledge; fourth, there is no evidence that these "unknowns" were a threat to the security of the country; and finally there was no physical or material evidence, not even a minute fragment, of a so-called "flying saucer" was ever found.

The Air Force emphasized the belief that if more immediate detailed objective observational data could have been obtained on the "unknowns" these too would have been satisfactorily explained.

A critical examination of the reports revealed that a high percentage of them were submitted by serious people, mystified by what they had seen and motivated by patriotic responsibility.

Reports of UFOs have aroused much interest on this subject throughout the country and a number of civilian clubs, committees and organizations have been formed to study or investigate air phenomena. These private organizations are not governmental agencies and do not reflect official opinion with respect to their theories or beliefs based upon observed phenomena or illusions.

No books, motion pictures, pamphlets, or other informational material on the subject of unidentified flying objects have been cleared, sponsored, or otherwise coordinated by the U. S. Air Force, with the exception of the official press releases issued by Headquarters, USAF, in Washington.

The Air Force, assigned the responsibility for the Air Defense of the United States, will continue to investigate, through the Air Defense Command, all reports of unusual aerial objects over the U.S., including objects that may become labeled Unidentified Flying Objects. The services of qualified scientists and technicians will continue to be utilized to investigate and analyze these reports, and periodic public statements will be made as warranted.

END

NEWS RELEASE
PLEASE NOTE DATE

DEPARTMENT OF DEFENSE
OFFICE OF PUBLIC AFFAIRS
Washington 25, D. C.

NO. 1108-57

NOVEMBER 15, 1957   LI 5-6700 Ext. 75131

FOR THE PRESS:

As a result of inquiries concerning Air Force evaluation of recnet Unidentified Flying Objects reports, the following general aspects of certain reports have been ascertained:

1.   LEVELLAND, Texas: (Big Light, seen by "dozens," stalled autos)

Investigation on the scene revealed that only three persons, rather than "dozens," could be located who saw the "big light." Preliminary reports have not revealed cause of "stalled" automobiles at this time, although rain and electrical storms at the time of the reported sightings, affecting wet electrical circuits, could be cause. Object visible only few seconds, not sustained visibility as had been implied.

EVALUATION: Weather phenomenon of electrical nature, generally classified as "Ball Lightning" or "St. Elmo's Fire," caused by stormy conditions in the area, including mist, rain, thunderstorms and lightning.

2.   ALAMAGORDO, New Mexico: (10 stalled autos, radio fade-out, "heat" light and "sun-burn")

Investigation of originator's report revealed no "sunburn" effect from "heat" light; originator admitted radio fade-out previously in same area; none of witnesses originator cited in other automobiles could be found after extensive search.

EVALUATION: Hoax, presumably suggested by the Levelland, Texas "reports."

3.   COAST GUARD CUTTER - Gulf of Mexico: (Object seen 3 seconds and radar contact)

Investigation based on data furnished by U. S. Coast Goard, indicated sighting four objects, three being radar and one visual, rather than one as has been widely reported. Radar returns on each of three contacts showed contact in almost direct opposite direction from preceding target, indicating spurious returns (false targets). However, a time-distance plot of two contacts indicated one target had approximate speed of 250 MPH and another a speed of 660 MPH, suggesting a propeller and jet type aircraft respectively. The one visual sighting for three seconds, described as bright, like a planet, moving at high speed, horizontally, also indicated high speed aircraft. The area of the sightings is the scene of many flights and operations by the Air Force and Navy, too numerous to check out based on these reports.

EVALUATION: Aircraft, and possible spurious radar returns.

4.   WHITE SANDS, New Mexico (Two patrols report UFO's separate times)

Investigation revealed that originators had discussed phenomena among themselves, accounting for similar stories, first given to local newspapers, then reported to their command. Astro plots indicate Venus is at magnitude at the time, place and direction of the first patrol's observation and the moon, with scattered clouds was in general direction of the second patrol's observation.

EVALUATION: Astronomical

5.   KEARNEY, Nebraska: (Space ship incident; stalled engines)

Investigation revealed that local officials consider originator wholly unreliable. However, another originator in same area who reported stalled engine in similar circumstances admitted having distributor rotor changed the day before and mechanic stated that after "stalling" incident, he found piece of old rotor wedged between points, which would cause engine failure.

EVALUATION: Hoax and engine failure.

END

NEWS RELEASE

PLEASE NOTE DATE

DEPARTMENT OF DEFENSE
OFFICE OF PUBLIC INFORMATION
Washington 25, D. C

IMMEDIATE RELEASE   October 6, 1958   NO. 986-58
                                        LI 5-6700 Ext. 75131

FACT SHEET

UNIDENTIFIED FLYING OBJECT "UNKNOWNS"
STILL BELOW 2 PERCENT IN LATEST 1270 CASES

According to latest information tabulated by Air Force authorities charged with investigation of Unidentified Flying Objects, 1270 new UFO reports were investigated during the period July 1 1957 through July 31, 1958. More than 84 percent of the reported UFO sightings were definitely established as natural phenomena, hoaxes, birds, or man-made objects. Insufficient data was available to thoroughly analyze and evaluate 14 percent of the reports and less than 2 percent were classied as unknowns.

Although investigations were completed on these "unknown" cases without positive or final conclusions, nothing was found to substantiate any claims that such sightings were interplanetary visitors, or so-called "Flying Saucers".

A sighting is considered unknown when the report contains sufficient information to draw at least one valid conclusion from the facts available but when the description of the reported object or its movements cannot be related to the norm. Refinements in Air Force investigative processes, however, have resulted in a steady decline in the number of UFO reports placed in this category. During the early months of the Air Force's investigation, some 20 percent of the reports were classified as unknown. These unknowns were reduced to 9 percent in 1953 and 1954, 3 percent in 1955, and 1.8 percent during the first six months of this year. It is believed that specific conclusions could be drawn for all reports if additional observational data were provided by individuals originating these reports. Also, the Air Force believes that there will always be a small number of unknowns due to high altitude phenomena which are strange to the untrained eye.

The 1270 new reports during the past thirteen months increases the total Air Force investigations conducted since June 1947 to more than 6,000.

All investigations of reported unusual aerial objects are conducted by the U.S. Air Force Air Defense Command personnel. These reports are reviewed by a select scientific advisory group at the Air Technical Intelligence Center which analyzes and evaluates, by scientific means, all information assembled by investigative teams.

MORE

Dr. J. Allen Hynek, Professor of Astrophysics and Astronomy at Ohio State University, is the chief scientific consultant to the Air Force on the subject of Unidentified Flying Objects.

His services and those of other eminently qualified civilian scientists are utilized if deemed necessary by the Air Force to further its investigation of individual cases or to perform detailed studies of the general subject. As an example, a panel of scientific consultants was called upon by the Air Force in 1953 to make an over-all examination of investigative procedures and findings on specific reports.

The panel said at that time "the evidence presented on UFOs showed no indication that these phenomena constitute a direct threat to national security." Further there was no "residuum of cases which indicates phenomena which are attributable to foreign artifacts capable of hostile acts." The panel concluded that "national security agencies should take immediate steps to strip the Unidentified Flying Objects of the special status they have been given and the aura of mystery they have unfortunately acquired."

In line with this recommendation, the Air Force periodically has made available through the national press and radio-television agencies, information on investigations and findings pertaining to cases which have received national attention. Only in a few instances were sighting reports classified to keep from compromising some elements in our Air Defense System, and no reports of sightings have been classified because they contain some authentic proof or even reasonable inferences that UFOs were real. Moreover, there are no Air Force documents which prove the existence of flying saucers or show that the earth is being observed, visited or threatened by machines from other planets.

However, because of limited personnel resources and to preclude possible embarrassment to individuals who have assisted Air Force investigators or who have originated reports later found to be erroneous, the Air Force does not honor individual requests for detailed analyses of specific cases.

A tabulation and evaluation of UFO report statistics follows:

1. The following represent the results of the evaluation of world-wide UFO reports investigated, processed and analyzed during the period July 1, 1957 through December 31, 1957, under pertinent categories:

   UFO Report Evaluations -- By Category and Percentage Breakdown:
   Total Reports 928

	Totals	Percent
Balloons	143	15.41
Aircraft	214	23.06
Astronomical	265	28.56
Others (Hoax, Lights, Birds, etc.)	152	16.38
Insufficient Data	139	14.98
Unknown	15	1.61
Sum Total	928	100.00

MORE

2. The following represents the results of the evaluation of world-wide UFO reports investigated, processed and analyzed during the period January 1, 1958 through July 31, 1958, under pertinent categories:

UFO Report Evaluations -- By Category and Percentage Breakdown:

Total Reports 342

	Totals	Percent
Balloons	51	14.9
Aircraft	76	22.2
Astronomical	89	26.0
Others (Hoax, Lights, Birds, etc.)	72	21.1
Insufficient Data	48	14.0
Unknown	6	1.8
Sum Total	342	100.0

3. UFO Evaluations -- Category Breakdown by Months: July 1957 - July 1958

	Balloons	A/C	Astro	Other	Insuf Data	Unknown
July	16	26	7	6	9	1
Aug.	18	29	27	9	17	1
Sept.	15	10	8	18	8	2
Oct.	25	19	37	17	13	3
Nov.	49	105	112	74	70	4
Dec.	20	25	74	28	22	4
Jan.	5	13	18	11	9	-
Feb.	7	11	15	9	5	1
March	9	3	13	13	8	1
April	7	17	12	10	8	1
May	7	8	9	7	4	1
June	5	8	9	8	4	1
July	11	16	13	14	10	1
Totals	194	290	354	224	187	21

E N D

NEWS RELEASE
PLEASE NOTE DATE

DEPARTMENT OF DEFENSE
OFFICE OF PUBLIC INFORMATION
Washington 25, D. C.

IMMEDIATE RELEASE   January 22, 1959   NO. 63-59
                                                                         LI 5-6700 Ext. 75131

AIR FORCE UFO STUDY
SHOWS "UNKNOWNS" DECREASING

The latest tabulation of Unidentified Flying Objects reports by Air Force authorities and scientists of the Air Technical Intelligence Center, covering the last half of 1958, has set a record low for the number of cases classified as "unknown", which is down to less than 1%.

The total number of sightings for the six month period was also lower than the number reported in the preceding half-year.

From July 1958 through December 1958, 296 cases of UFO reports were analyzed. This was a 14% decrease in sightings over the first half of the year.

As in previous compilations, the astronomical category accounted for the highest percentage of sightings at 32% of the total. A new category, satellites, accounted for only 1.36% and the "unknown" category amounted to only 0.67%.

A statistical tabulation of the six months period follows:

UFO Report Evaluations -- By Category and Percentage Breakdown:

Total Reports 296

	Totals	Percent
Balloons	23	7.77
Aircraft	60	20.27
Astronomical	96	32.44
Satellite	4	1.36
Other	50	16.89
Insufficient Data	61	20.60
Unknown	2	.67
Sum Total	296	100.00

MORE

UFO Evaluations -- Category Breakdown by Months:
July 1958 - December 1958

	Balloons	A/C	Astro	Satellite	Other	Insuf Data	Unknown
July	11	16	13	0	14	10	1
Aug	5	17	20	2	14	19	0
Sep	1	10	19	0	8	12	0
Oct	3	11	19	1	10	8	1
Nov	2	5	11	0	2	7	0
Dec	1	1	14	1	2	5	0
Totals	23	60	96	4	50	61	2

E N D

NEWS RELEASE
PLEASE NOTE DATE

DEPARTMENT OF DEFENSE
OFFICE OF PUBLIC AFFAIRS
Washington 25, D. C.

IMMEDIATE RELEASE   July 15, 1959   NO. 798-59
                                                OXford 7-5131

AIR FORCE REPORTS UNIDENTIFIED FLYING OBJECT SIGHTINGS DECREASE BY FIFTY PERCENT

Over twelve years of investigating and evaluating unidentified flying object sightings has provided no evidence to confirm the existence of the popularly termed "flying saucers", as interplanetary or interstellar space ships. The total number of sightings for the first six months of 1959 was only 143. This is in sharp contrast to the previous six months total of 296 or a little more than a 50 percent drop in reported sightings.

As in previous compilations, the astronomical category accounted for the highest percentage of sightings at 45 percent of the total. The unidentified category amounted to only three cases or 2 percent of the total sightings for the period.

A statistical tabulation for the six months period follows:

UFO Report Evaluation - By Category and Percentage Breakdown

Balloons	7	4.89%
Aircraft	23	16.09%
Astronomical	65	45.45%
Other (Birds, Hoaxes, Searchlights, etc.)	19	13.29%
Insufficient Data	26	18.19%
Unidentified	3	2.09%
Total Reports	143	

UFO Evaluations - Category Breakdown by Months 1 January 1959 - 30 June 1959

	Balloons	A/C	ASTRO	OTHER	INSUF DATA	UNIDEN-TIFIED	TOTAL	%
Jan	0	3	15	4	5	0	27	16.88
Feb	3	5	12	2	5	1	28	19.58
Mar	0	3	15	4	3	1	26	18.18
Apr	1	4	10	4	3	1	23	16.08
May	3	3	7	4	4	0	21	14.69
June	0	5	6	1	6	0	18	12.59
Total	7	23	65	19	26	3	143	
%	4.89	16.09	45.45	13.29	18.19	2.09	-	100

The Air Force, assigned the responsibility for the Air Defense of the United States, will continue to investigate all reports of unusual aerial sightings over the United States, including objects that may become labeled Unidentified Flying Objects. The services of qualified scientists and technicians will continue to be utilized to investigate and analyze reports, and periodic public statements and releases will be made as warranted.

E N D

NEWS RELEASE
PLEASE NOTE DATE

DEPARTMENT OF DEFENSE
OFFICE OF PUBLIC AFFAIRS
Washington 25, D. C.

IMMEDIATE RELEASE   JANUARY 29, 1960   NO. 98-60
                                                                       OXford 7-5131

FACT SHEET
AIR FORCE UFO REPORT

After thirteen years of investigating and evaluating unidentified flying object sightings, the United States Air Force has found no evidence to confirm the existence of the popularly termed "flying saucers" as interplanetary or interstellar space ships. Total numbers of sightings reported to the Air Force over the thirteen year period are as follows:

YEAR	Objects sighted and reported
1947	79
1948	143
1949	186
1950	169
1951	121
1952	1501
1953	425
1954	429
1955	404
1956	778
1957	1178
1958	573
1959	326
	6312

It appears that some specific incident is usually responsible for touching off a rash of reported sightings and this is particularly true of the two high report years of 1952 and 1957.

In 1952 the Washington, D. C. sightings with extensive national press coverage resulted in 1501 sightings reported that year.

In 1957, the Russian Satellite Sputnik I was launched in October and particularly significant is the fact that 701 of the total of 1178 sightings reported were made in the last three months of that year.

During the period 1 July 1959 through 31 December 1959, there were 183 cases of unidentified aerial phenomena reported to the Air Force as compared to 310 cases for the same period in 1958

MORE

This represents a reduction of 41 percent in the number of cases reported. Of t' 183 cases reported to the Air Force, 167 have been analyzed and evaluated, and 16 cas are pending awaiting additional data. UFO cases or files are not closed and should additional information be obtained at a later date, it can result in a solved case and/ or may change the category previously assigned to the individual sighting (as in the case of a previously unidentified case being changed to the aircraft category). This open end policy on all UFO files results in minor changes from time to time in UFO category statistics.

UFO Report Evaluation - by Category and Percentage Breakdown
1 July 1959 - 31 December 1959

Category	Number	Percent
Astronomical	39	23.34
Aircraft	16	9.59
Balloon	10	5.99
*Insufficient Data	70	41.92
Other (Birds, hoaxes, searchlights, etc.)	28	16.76
Unidentified	4	2.40
	167	100%

*This category includes 37 sightings (22.16%) reported by individuals and in each case there was no supporting witness and no additional information or facts available to allow a scientific evaluation of the sighting. Sightings of this nature must be discounted because of a total lack of supporting data. This in effect reduces the total number of cases for calendar year 1959 to only 146 with 130 of these evaluated to date. In the case of the other 33 sightings in the insufficient data category, additional data or facts were available but were not sufficient enough to allow a valid conclusion.

Reporting, investigation, analysis and evaluation procedures have improved considerably since the first sighting of a "flying saucer" was made on 27 June 1947. The study and analysis of reported sightings of UFO's is conducted by a selected scientific group under the supervision of the Air Force.

Dr. J. Allen Hynek, Professor of Astrophysics and Astronomy at Northwestern University, is the Chief Scientific Consultant to the Air Force on the subject of Unidentified Flying Objects.

The selected, qualified scientists, engineers, and other personnel involved in these analyses are completely objective and open minded on the subject of "flying saucers." They apply scientific methods of examination to all cases in reaching their conclusions. The attempted identification of the phenomenon observed is generally derived from human impressions and interpretations and not from scientific devices or measurements. The data in the sightings reported are almost invariably subjective in nature. However, no report is considered unsuitable for study and categorization and no lack of valid evidence of physical matter in the case studies is assumed to be "prima facie" evidence that so-called "flying saucers" or interplanetary vehicles do not exist.

General categories of identification are balloons, aircraft, astronomical, other, insufficient data, and unidentified.

Approximately 4,000 balloons are released in the U.S. every day. There are two general types of balloons: weather balloons and upper air research balloons. Balloons will vary from types 4 feet in diameter to large types 200 feet in diameter. The majority released at night carry running lights which often contribute to weird or unusual appearances when observed at night. This also holds true when observed near dawn or sunset because of the effect of the slant rays of the sun upon the balloon surfaces. The large balloons, if caught in jet streams, may assume a near horizontal position when partially inflated, and move with speeds of over 200 MPH. Large types may be observed flattened on top. The effect of the latter two conditions can be startling even to experienced pilots.

Many modern aircraft, particularly swept and delta wing types, under adverse weather and sighting conditions are reported as unusual objects and "flying saucers." When observed at high altitudes, reflecting sunlight off their surfaces, or when only their jet exhausts are visible at night, aircraft can have appearances ranging from disc to rocket in shape. Single jet bombers having multi-jet pods under their swept-back wings have been reported as UFO's or "saucers" in "V" formation. Vapor trails will often appear to glow with fiery red or orange streaks when reflecting sunlight. After burners are frequently reported as UFOs.

The astronomical category includes bright stars, planets, comets, meteors, and other celestial bodies. When observed through haze, light fog, or moving clouds, the planets Venus, Mars, and Jupiter have often been reported as unconventional, moving objects. Attempts to observe astronomical bodies through hand-held binoculars under adverse sky conditions has been a source of many UFO reports.

The "other" category includes reflections, searchlights, birds, kites, blimps, clouds, sun-dogs, spurious radar indications, hoaxes, firework displays, flares, fireballs, ice crystals, bolides, etc., as examples: large Canadian geese flying low over a city at night with street lights reflecting off their bodies; searchlights playing on scattered clouds, appearing as moving disc-like shapes.

The insufficient data category includes all sightings where essential or pertinent items of information are missing, making it impossible to form a valid conclusion. These include corroboration of the sighting by an additional witness; description of the size, shape or color of the object; direction and altitude; exact time and location; wind weather conditions, etc. This category is not used as a convenient way to get rid of what might be referred to as "unidentified objects." However, if there is not an additional witness or if the data received is insufficient or unrelated, the analysts must then place that particular report in this category. The Air Force needs complete information to reach a valid conclusion. Air Force officials stressed the fact that an observer should send a complete report of a bona fide sighting to the nearest Air Force activity. There the report will be promptly forwarded to the proper office for analysis and evaluation.

A sighting is considered unidentified when a report contains all pertinent data necessary to normally suggest at least one valid hypothesis on the cause or explanation of the sighting but when the description of the object and its maneuvers cannot be correlated with any known object or phenomenon. In its Project Blue Book Special Report #14, released in October 1955, the Air Force showed that evaluated sightings in the unidentified category had been reduced to 3 percent at that time.

MORE

Unidentified sightings had been 9 percent in 1953 and 1954 and in previous years unidentified sightings had run as high as 20 percent. Project Blue Book Special Report #14, covered "flying saucer" investigations from June 1947 to May 1955. Since that time, Air Force statistics show the percentage of unidentified sightings has been reduced to approximately 2 percent.

Air Force conclusions for the thirteen years of UFO sightings involving over 6,000 reports are: first, there is no evidence that unidentified sightings were inimical or hostile; second, there is no evidence that these unidentified sightings were interplanetary space ships; third, there is no evidence that these unidentified sightings represented technological developments or principles outside the range of our present day scientific knowledge; fourth, there is no evidence that these unidentified sightings are a threat to the security of the country; and finally, no physical or material evidence, not even a minute fragment of a so-called "flying saucer" has ever been found.

The Air Force emphasized the belief that if more immediate detailed objective observational data could have been obtained on the unidentified sightings, these, too, would have been explained satisfactorily.

The Air Force, assigned the responsibility for the Air Defense of the United States, will continue to investigate all reports of unusual aerial objects over the U. S. including objects that may become labeled Unidentified Flying Objects. The services of qualified scientists and technicians will continue to be utilized to investigate and analyze these reports, and periodic press releases on the subject will be made as warranted.

END

NEWS RELEASE
PLEASE NOTE DATE

DEPARTMENT OF DEFENSE
OFFICE OF PUBLIC AFFAIRS
Washington 25, D. C.

IMMEDIATE RELEASE   July 21, 1960   NO. 812-60
                                          OXford 75131

FACT SHEET
AIR FORCE UFO REPORT

A total of 173 sightings of unidentified flying objects or aerial phenomena were reported to the USAF during the period 1 January to 30 June 1960. Of these 173 cases 139 have been analyzed and 34 are pending.

Of the 139 cases analyzed 51 were categorized "Insufficient Evidence." Of these 51 cases 37 were so categorized because there were no additional witnesses. It is the Air Force policy to attempt to determine the probable cause of these sightings. However in keeping with good scientific practices they are categorized for statistical purposes as "insufficient evidence." To give full credit to these particular sightings would be comparable to accepting as fact the results of an experiment which was conducted only once.

During this same period for 1959 the number of sightings reported to the Air Force was 175. Due to the fact that several reports usually reach the Air Force after the end of a period, it is safe to assume that the 1 Jan - 30 June period of 1960 will slightly exceed that of 1959. However, the 173 cases for the first half of 1960 represents a slight decrease from the 189 cases reported to the Air Force for the period 1 July through 31 December 1959.

During the period of this report there were 41 cases reported from foreign countries. The majority of these sightings took place in the Pacific and Far East area.

It is apparent that the public is taking the reporting of unidentified aerial phenomena more seriously. During this reporting period there was only one instance of an actual hoax.

The months of April and June 1960 show a fairly large percentage of astronomical sightings. These large percentages can be attributed to the excessive meteor activity in April and to the proximity of Jupiter in June. Its nearness resulted in an apparent brightness greater than that of a star of -2.0 magnitude and the planet also appeared unusually large.

When categorizing sightings of unidentified aerial phenomena the Air Force attempts to do so primarily by association to a responsible object. If the planet Venus was determined to have been the subject of a report the case is categorized as an astronomical sighting. It should be noted that in most instances the condition under which an object is seen is the real cause of the sighting and not the object itself. An example of this would be a star or planet seen low on the horizon with the associated refraction and diffraction patterns. This brings to point the secondary method of categorizing sightings. In some instances the evidence in a case will indicate the probable cause as a reflection, etc., but the responsible object may not be as traceable as a

MORE

planet or star. In these instances the cases are categorized according to probably cause. A third method is the categorizing of those cases which are determined to be illusions, probably resulting from over active imaginations. Usually these cases are one witness sightings and ultimately fall into the "insufficient evidence" category.

There are still many sightings which are due to the viewing of startling natural objects for the first time; i.e., fireball type meteors. In most of these instances the object itself, coupled with the relative inexperience of the witness, is the resulting cause of the sighting.

It is significant to note the adaptation of the public to the satellite era of the space age as evidenced by only one satellite being reported as an UFO during this period. The one satellite reported was the startling re-entry of the Discoverer VIII on 7 March 1960.

To date no report of unidentified aerial phenomena gave any indication of a threat to the security of the United States, nor were there indications that these phenomena were other than natural.

UFO Report Evaluation - by Category and Percentage Breakdown

1 January 1960 - 30 June 1960

Astronomical	52	37.41
Aircraft	7	5.04
Balloon	6	4.31
*Insufficient Data	51	36.70
Other (Birds, hoaxes, search lights, etc.)	19	13.67
Satellites (Re-entry of Discoverer VIII)	1	.72
Unidentified	3	2.15
	173	100%

*This category includes 37 sightings (21.38%) reported by individuals and in each case there were no supporting witness and no additional information or facts available to allow a valid scientific evaluation of the sighting. Sightings of this nature must be discounted because of a total lack of supporting data. This in effect reduces the total number of cases for the period to only 136 with 102 of these evaluated to date. In the case of the other 14 sightings in the insufficient data category, addtional witnesses corobrated with sightings but did not provide sufficient data to allow a valid conclusion. Percentages are straight mathematical computations carried out to only two places based upon totals and sub totals or categories. Therefore when worked in reverse they will not reflect the exact original total or sub total.

UFO cases or files are not closed and should additional information be obtained at a later date, it can result in a solved case and/or may change the category previously assigned to the individual sighting (as in the case of a previously unidentified case being changed to the aircraft category). This open and policy on all UFO files results in minor changes from time to time in UFO category statistics. This also applies to individual reports submitted after the fact. These continuing inputs account for additional minor changes periodically which, of course, are not consistent with previously released figures.

The total number of sightings reported to the United States Air Force since 1947 are as follows:

YEAR	OBJECTS SIGHTED AND REPORTED
1947	79
1948	143
1949	186
1950	169
1951	121
1952	1501
1953	425
1954	429
1955	404
1956	778
1957	1178
1958	573
1959	364
1960 (through 30 June)	173
	6523

Reporting, investigation, analysis and evaluation procedures have improved considerable since the first sighting of unidentified flying object was made on 27 June 1947. The study and analysis of reported sightings of UFO's is conducted by a selected scientific group under the supervision of the Air Force.

Dr. J. Allen Hynek, Head of the Department of Astronomy and Director of the Observatory at Northwestern University, is the Chief, Scientific Consultant to the Air Force on the subject of Unidentified Flying Objects.

The selected, qualified scientists, engineers, and other personnel involved in these analyses are completely objective and open minded on the subject of UFO's. They apply scientific methods of examination to all cases in reaching their conclusions. The attempted identification of the phenomenon observed generally must be derived from human impressions and interpretations because scientific devices or measurements are not available. The data in the sightings reported are almost invariably subjective in nature. However, no report is considered unsuitable for study and categorization.

General categories of identification are balloons, aircraft, astronomical, other, insufficient data, satellites and unidentified.

Approximately 4,000 balloons are released in the U. S. every day. There are two general types of balloons: weather balloons and upper air research balloons. Balloons will vary from types 4 feet in diameter to large types 200 feet in diameter. The majority released at night carry running lights which often contribute to weird or unusual appearances when observed at night. This also hold true when observed near dawn or sunset because of the effect of the slant rays of the sun upon the balloon surfaces. The large balloons, if caught in jet streams, may assume a near horizontal position when partially inflated, and move with speeds of over 200 MPH. Large types may be observed flattened on top. The effect of the latter two conditions can be startling even to experienced pilots.

Many modern aircraft, particularly swept and delta wing types, under adverse weather and sighting conditions are reported as unusual objects and/or "flying saucers." When observed at high altitudes, reflecting sunlight off their surfaces, or when only their jet exhausts are visible at night, aircraft can have appearances ranging from disc to rocket in shape. Single jet bombers

-3-

MORE

having multi-jet pods under their swept-back wings have been reported as UFOs or "saucers" in "V" formation. Vapor trails will often appear to glow with fiery red or orange streaks when reflecting sunlight. After burners are frequently reported as UFOs.

The astronomical category includes bright stars, planets, comets, meteors, and other celestial bodies. When observed through haze, light fog, or moving clouds, the planets Venus, Mars, and Jupiter have often been reported as unconventional, moving objects. Attempts to observe astronomical bodies through hand-held binoculars under adverse sky conditions have been a source of many UFO reports.

The "other" category includes reflections, searchlights, birds, kites, blimps, clouds, sun-dogs, spurious radar indications, hoaxes, firework displays, flares, fireballs, ice crystals, bolides, etc., as examples: large Canadian geese flying low over a city at night with street lights reflecting off their bodies; searchlights playing on scattered clouds, appearing as moving disc-like shapres.

The insufficient data category includes all sightings where essential or pertinent items of information are missing, making it impossible to form a valid conclusion. These include corroboration of the sighting by an additional witness; description of the size, shape or color of the object; direction and altitude; exact time and location; wind weather conditions, etc. This category is not used as a convenient way to get rid of what might be referred to as "unidentified objects." However, if there is not an additional witness or if the data received is insufficient or unrelated, the analysts must then place that particular report in this category. The Air Force needs complete information to reach a valid conclusion. Air Force officials stressed the fact that an observer should send a complete report of a bona fide sighting to the nearest Air Force activity. There the report will be promptly forwarded to the proper office for analysis and evaluation.

A sighting is considered unidentified or unexplained when a report apparently contains all the pertinent data necessary to normally suggest at least one valid hypothesis on the cause or explanation of the sighting but when the description of the object and its maneuvers cannot be correlated with any known object or phenomenon.

In its Project Blue Book Special Report #14, released in October 1955, the Air Force showed that evaluated sightings in the unidentified category had been reduced to 3 percent at that time.

Unidentified sightings had been 9 percent in 1953 and 1954 and in previous years unidentified sightings had run as high as 20 percent. Project Blue Book Special Report #14, covered UFO investigations from June 1947 to May 1955. Since that time, Air Force statistics show the percentage of unidentified sightings has been reduced to approximately 2 percent.

Air Force conclusions for the thirteen years of UFO sightings involving over 6,500 reports are: first, there is no evidence that unidentified sightings were inimical or hostile; second, there is no evidence that unidentified sightings were interplanetary space ships; third, there is no evidence that these unidentified sightings represented technological developments or principles outside the range of our present day scientific knowledge; fourth, there is no evidence that these unidentified sightings are a threat to the security of the country; and finally, no physical or material evidence, not even a minute fragment of a so-called "flying saucer" or space ship has evern been found.

The Air Force emphasized the belief that if more immediate detailed objective observational data could have been obtained on the unidentified or unexplained sightings, these, too, would have been explained satisfactorily.

The Air Force, assigned the responsibility for the Air Defense of the United States, will continue to investigate all reports of unusual serial objects over the U. S. including objects that may become labeled Unidentified Flying Objects. The services of qualified scientists and technicians will continue to be utilized to investigate and analyze these reports, and periodic public statements on the subject will be made as warranted.

The Air Force Inspector General's Brief, dated 24 December 1959, contained a notice to all unit Commanders that UFO reports are serious business since they are vitally involved in the Air Force's air defense mission. This Brief stressed that Commanders are responsible for seeing that UFO sightings are investigated and evaluated quickly, thoroughly and accurately. The Brief referred to Air Force Regulation 200-2 concerning unidentified flying objects which outlines the Air Force's obligation to keep the public adequately informed on this subject and also stresses the fact that UFO sighting reports are not to be classified. If an UFO report is classified it is classified for other reasons than the UFO sighting itself.

The Air Force is charged by many private UFO groups with possession of classified information which concludes or proves that space ships from other planets exist and are visiting our atmosphere. Nothing could be further from the truth. The Air Force possesses no information, classified or unclassified, which proves this contention. Many individuals associated with these private UFO organizations are self-appointed authorities on UFOs and consider themselves entitled to be unofficial advisors to the United States Air Force Intelligence Community. Since they are not charged under law with these responsibilities it would be entirely unappropriate and even dangerous at times to exercise the Intelligence system in order to give them, or their organizations, any notoriety or publicity. It is an interesting fact that most of these individuals fail to recognize the fact that there is such a thing as legitmate classified security information. For the past thirteen years these groups have repeatedly stated that their organizations possess documented evidence proving the existence of space ships from other planets. However, when asked as patriotic citizens to deliver this evidence to the United States Air Force or other government agencies they have refused to do so. The obvious conclusion is that they do not possess this information or evidence proving the existence of space ships.

From time to time the Air Force has been asked the cost of conducting the UFO program. It is impossible to estimate the exact amount of funds expended by the Air Force in investigating reported sightings of unidentified flying objects because in addition to the entire facilities of the Aerospace Technical Intelligence Center, the entire facilities of the nation's scientific and technical community are at the disposal of the evaluators, and the degree of use varies with each case. Also, investigation in the field can involve individual Air Force base personnel and special investigative units and again, in these instances the amount of time expended and the number of personnel assigned to a specific sighting will vary with each case.

END

INDEX

Accident and Investigation Board, 20
Adamski, George, 38, 41
Aerospace Technical Intelligence Center, 5, 6, 8, 10, 11, 14, 18, 19, 20, 21, 22, 23, 24, 25, 84, 85
  alleged documents of, 33, 83
after-image, 68
air cars, 87
Alpha Centauri, 40
Alpha Orionis, 12
American Meteor Society, 8
Arnold, Kenneth, 12, 13, 14, 15, 58, 84
auroras, 64
autohypnosis, 69
autokinesis, 69

balloons, confusion with UFO's, 53

Calder, Dr. William A., 78
Central Intelligence Agency, "report" of, 33
cirrus, 63
Civil Aeronautics Authority Air Route Traffic Control Center, 21, 24
comets, 61-63
  and after-image, 68; faint, 80

Defense, Department of, 87; Congressional inquiries to, 8; press release on UFO's, 38, 41
Dew Line Extension, 56
Discoverer, 57
  satellite carrier of, 29
distance, judgment of, 72
Doolittle, General James H., 77
Dryden, Dr. Hugh L., 77
Duval, Dr. Addison M., 77

effects, physiological, on observation, 68
Elvey, Dr. Christian, 10
eye, movements of the, 72

fatigue, 69
  physiological changes due to, 72
fear of another world war, 46
fear of the unknown, 32, 37
Federal Bureau of Investigation, 16
Flying Saucer Conspiracy, The, by Major Keyhoe, 31, 38
Flying Saucers, by Donald Menzel, 79
Flying Saucers Have Landed, by George Adamski, 41

Grenstein, Dr. Jesse L., 80
Ground Observer Corps, 9

Halley's Comet, 61
halo, 61, 63

Hoffmeister, Professor Guno, 8
Horner, Assistant Secretary Richard F., 76, 77
House of Representatives
  Armed Services Committee, 50; Committee on Science and Astronomy, 50; Committee on Space and Aeronautics, 50
Hynek, Professor J. A., 15, 78, 85
hypoxia, 68

Inside the Spaceships, by George Adamski, 41
International Geophysical Year, 76

Jupiter, 60
  orbit of, 61; stellar magnitude of, 61

Kamp, Dr. Peter van de, 77
Keyhoe, Major Donald E., 31, 36, 38, 39, 40

Ley, Willy, 78
life on other planets, 32, 74
Long Enduring Meteor Trains and Fireball Orbits, by Charles P. Olivier, 7

Mars, 17, 32, 34, 39, 42
  confusion of with UFO's, 55; orbit of, 61; "radio waves" from, 37; stellar magnitude of, 60; "water on," 41
Martians, contact with claimed, 30-31
"Master Aetherius," 43
Menzel, Dr. Donald, 20, 74, 75, 79
Mercury
  orbit of, 60; phases of, 61; stellar magnitude of, 60
meteor(s), 10, 62-63, 80
  and after-image, 68; bolide, 11, 23, 62; confusion of with UFO's, 51, 58; fireballs, 8, 23, 26, 54, 62, 63, 79, 80; identification of, 53-54; radar response to, 67; and sightings, 28; trails of, 8, 81
mirage(s), 79
  causes of, 56, 58-59, 75
missiles, 7, 10
moon dogs, 63, 64
movement, autokinetic, 72

National Advisory Committee for Aeronautics, 77
National Aeronautics and Space Administration, 77

Olivier, Charles P., 7, 8

paraselene, 63
parhelia, 63

perception, 70-71
Pioneer V, 81

Quarles, Secretary Donald A., 17

radar echoes
 production of, 65-67; responses, 56
Rand Corporation, 16
Report on Unidentified Flying Objects, The, by Captain Ed Ruppelt, 77

St. Elmo's fire, 12
satellites, 7, 10
Satellites, Rockets, and Outer Space, by Willy Ley, 78
saucers, flying, 9, 13, 17, 23, 29, 31, 32, 41, 42, 46, 75, 80
 believers in, 32; clubs, 47, 49; epidemic of, 78; evidence of, 78; explained as natural occurrences, 76; extra-terrestrial origin of, 77; and flying shields, 12, 16; literature, 79; men on, 32; organizations interested in, 47; psychology of belief in, 30; "real," 79, 83; and sense of sight, 68; suggestion that USAF develop, 33; development of, 87
"saucerian" or "contact groups," 46
Senate of the United States
 Committee on Space and Aeronautics, 50; Committee on Space and Astronautics, 50; Preparedness Subcommittee of the Senate Armed Services Committee, 50; Subcommittee on Government Operations, 50
sense of sight, 68-70
sighting(s)
 Boise, 15; explanations of specific, 23-25, 27, 28; Montgomery, 20-21; Mt. Rainier, 12-13, 14; Muroc, 15; Nome-Unalakleet, 10; Pacific, 8, 10, 10n; Portland, 15; production of radar responses by, 56; psychology of, 19
Sirius, 61, 75
space people, belief in, 46
 no proof of offered, 37
space ships, 9, 16, 20, 30, 31, 41, 46, 47, 74
 interplanetary, 17, 81, 83; no proof of offered, 37
Sputnik I, 83
Strategic Air Command, 13
Struve, Dr. Otto, 77
sub-suns, 63
sun dogs, 63, 64

temperature inversion
 influence of on radar, 67; mirages caused by, 56, 59; sightings due to, 22

Truth About Unidentified Objects, The, by Greene-Rouse, 42

"UFO, The Enigma of the Skies," television presentation, 74
UFO's. *see* unidentified flying object(s)
"ufologists," 46
unidentified flying object(s), 10, 11, 16, 17, 31, 39, 57, 75
 analysis of maneuvers of, 83; "ATIC document" on, 33; belief in, 30, 40; books, 86; cause of sightings of, 70; "classics," 19; clubs, 47, 49, 50; comets and meteors confused with, 61, 68; confusion of with familiar objects, 51; defense against, 87; definition of, 51; error in judgment in seeing, 73; information on, 76; "interplanetary origin" of, 36; interest in, 46; interest of Congressional Committees in, 47; "investigators and researchers" of, 34; "journals," 34; and "light energy," 38; number of sightings and reports of, 72; and optical illusions, 58; and picket ships, 56; press releases on, 18; program, 82, 83, 84, 85; reports on, 74; and satellite launchings, 57; sightings of finally solved, 29; variety of shapes of, 54-55
United States Air Force, 10n, 11, 13, 15, 16, 17, 24, 27, 78, 79, 86, 87
 attitude of, 47; briefings to Congress by, 50; correspondence received by, 31, 83; examination of sightings, 46; hoax reported to, 48; Information Office, 30; Intelligence Section, 22; investigation of UFO's by, 82; Press Desk, 6, 21; "Project Bluebook," 14, 17, 18, 75, 77; report on UFO's, 18; Scientific Advisory Board of, 16, 17, 84; School of Aerospace Medicine, 69; UFO's reported to, 74; withholding of information charged to, 33, 35, 37, 40, 47, 76, 86

Venus, 17, 20, 43
 confusion of with UFO's, 27, 54, 55; distortion of, 59; "message" from, 46; phases of, 61; "radio waves" from, 37; stellar magnitude of, 60; "water on," 41
Venusians, contact with claimed, 30-31
vision, stare, 72

"Washington Saucer Intelligence," 49
White, General Thomas D., 82

SPACE AGE READING

PUBLISHED BY

Van Nostrand

✳

ROCKET TO THE MOON
by Eric Bergaust and Seabrook Hull

✳

ROCKETRY AND SPACE EXPLORATION: The International Story
by Andrew G. Haley

✳

SPACE AGE DICTIONARY
by Charles McLaughlin

✳

SPACE FLIGHT
by Krafft A. Ehricke

✳

VAN NOSTRAND'S DICTIONARY OF GUIDED MISSILES and SPACE FLIGHT
Edited by Grayson Merrill, Captain, U.S.N. (Ret.)

Veteran of 15 combat missions and two battle campaigns as an 8th Air Force Navigator in European Theater of Operations, Lieutenant Colonel Lawrence J. Tacker, 43 years old, of Falls Church, Virginia, is now assigned as Chief of the Magazine and Book Branch, Office of Information, Office of the Secretary of the Air Force, Washington 25, D. C.

Colonel Lawrence Tacker with approximately 4,000 flying hours is rated as a Master Navigator with the United States Air Force and is presently fully qualified in the latest Air Force multi-jet engine aircraft. He holds the Air Medal, the Purple Heart, the European Theater of Operations Medal with Battle Stars, the Army and the Air Force Commendation Ribbons and has 19 years service with the U.S.

He, his wife, Dorothy and their four children: Francine Mary, age 14; Thomas Wood, 12; John Lawrence, 10; and Timothy Francis, 7 reside in an old colonial house on two acres of ground in historic and beautiful northern Virginia. By necessity, as well as inclination, Colonel Tacker's hobby interests remodelling and gardening.

www.ingramcontent.com/pod-product-compliance
Lightning Source LLC
Chambersburg PA
CBHW071232170426
43191CB00032B/1358